Fluorite in Utah

U.S. Geological Survey Bulletin 1005

*Compiled by the staff of
the United States Geological Survey*

with an introduction by Kerby Jackson

*This work contains material that was originally published in 1954 by
the United States Geological Survey, in part with assistance from the U.S. Atomic Energy
Commission and the State of Utah.*

COVER CREDITS -

Cover Design :
Gold Rush Books, Oregon, USA.

Cover Photograph :

Deep green isolated fluorite crystal showing cubic {100} and octahedral {111} faces, complete and undamaged, set upon a micaceous matrix. Overall size: 50 mm x 27 mm. Crystal size: 19 mm wide. Weight: 30 g. From Erongo Mountain, Erongo Region, Namibia.

Source: CarlesMillan, Wikipedia

Introduction

It has been sixty years since the United States Department of Geological Survey released its classic work, "Fluorspar Deposits in Utah".

Fluorite, also commonly referred to as "fluorspar" is a hallide mineral that typically crystallizes in an isometric cubic habit. This very colorful gem is the mineral form of calcium fluoride and is treasured by mineral collectors, as well as those who appreciate semi-precious stones. While the mineral also has industrial uses as a flux in iron smelting, in the production of hydrofluoric acid and also in the manufacture of special optics, but it is best known among the public for its ornamental value and is widely collected by those interested in gemstones and minerals.

Typically, fluorite occurs in hydrothermal veins with metallic minerals, where it often forms as part of the gangue and is usually associated with galena, sphalerite, barite, calcite and quartz.

Fluorite naturally occurs in a wide range of colors due to the fact that it is allochromatic, meaning that it can be tinted in color by elemental impurities. While the most common colors include purple, blue, green, yellow and clear, fluorite in hues of pink, red, white, brown and black are also sometimes found.

In 1852, George Gabriel Stokes coined the term "fluoresecence" after fluorite due to its ability to emit a peculiar glow when placed under a source of ultraviolet light. Usually, fluorite emits a blue light when so exposed, but some specimens also emit red, purple, yellow, green and white. This variation in hue is usually associated with the sample's place of origin, but while the impurities in the fluorite, such as the elements yttrium, ytterbium and others may effect the color of light emitted, the color of the fluorescence is not considered a guideline to determining the type of impurities found in a specimen, nor the stone's origin. Fluorite also possesses properties of thermoluminescence which is the ability to emit light as a result of previous exposure to electromagnetic radiation when the specimen is exposed to heat.

Today, there is a great interest in prospecting for fluorite and other gem minerals and these new prospectors are hungry for information on where the treasure they are seeking has been previously found, knowing full well that the easiest way to find a particular mineral, is to look in the same area where that mineral was previously found and to rely on the clues that their forerunners left behind.

This important volume and others like it, are being presented in their entirety again, in the hope that the average prospector will no longer stumble through the overgrown hills and the tailing strewn creeks without being well informed enough to have a chance to succeed at his ventures.

Kerby Jackson
Josephine County, Oregon
May 2014

FLUORSPAR DEPOSITS OF UTAH

By W. R. Thurston, M. H. Staatz, D. C. Cox, and others

ABSTRACT

The studies of fluorspar localities in Utah made by the U. S. Geological Survey during and since the recent war are summarized. The fluorspar at the Cougar Spar and Blue Bell mines in the Indian Peak Range of western Beaver County occurs as fissure veins in fault and breccia zones in volcanic and intrusive rocks. At the Monarch (Staats) claims in west-central Beaver County fluorspar was mined chiefly from a fault between limestone and rhyolite porphyry. The Thomas Range district in Juab County has yielded sizeable tonnages of fluorspar from pipes in faulted dolomite and rhyolite porphyry. From 1918 to 1924 the Silver Queen mine in Tooele County produced fluorspar from fissure veins in faulted limestone.

The report describes the geology of producing mines and the various prospects examined. Production and reserves of fluorspar for Utah are summarized.

INTRODUCTION

By W. R. Thurston

Fluorspar has been reported from many parts of Utah (fig. 1), and in recent years the State's production has become increasingly important. The annual fluorspar production (fig. 2) increased from an average of less than 500 tons for the period from 1935 through 1943 to about 9,500 tons in 1948, 8,300 tons in 1949, and 18,900 tons in 1950. This trend toward expanding production can be expected to continue.

Production of commercial quantities has been limited to Beaver, Juab, and Tooele Counties. In Beaver County fluorspar has been mined in the Star district near Milford, in the Monarch (Staats) area 47 miles west of Milford, and in the Indian Peak Range near the Utah-Nevada State line. In Juab County the fluorspar area is in the Thomas Range in the center of the county, about 48 miles east of the Utah-Nevada State line. The fluorspar mined in Tooele County came from the Silver Queen (Wildcat) mine near Clive. Although the tonnage of ore mined from these three areas has not formed a large part of the national production (fig. 2), it did contribute materially to the fluorspar needed during the recent war. Minor occurrences of fluorite are listed in table 3.

The known reserves of fluorspar in Utah total about 450,000 tons of material having a minimum of 40 percent of CaF_2. Further

1

FIGURE 1.—Index map of Utah showing distribution of fluorspar deposits.

geologic study and prospecting are needed, however, before the potentialities of most of the fluorspar deposits of Utah are known.

This report summarizes the studies of fluorspar in Utah that were made as a part of the recent wartime program for the investigation of strategic and critical minerals. The work consisted of detailed geologic mapping of individual deposits to obtain information on the occurrence and distribution of minable fluorspar and to aid in exploring and developing fluorspar during World War II. Between August 1945 and July 1946, the investigations were

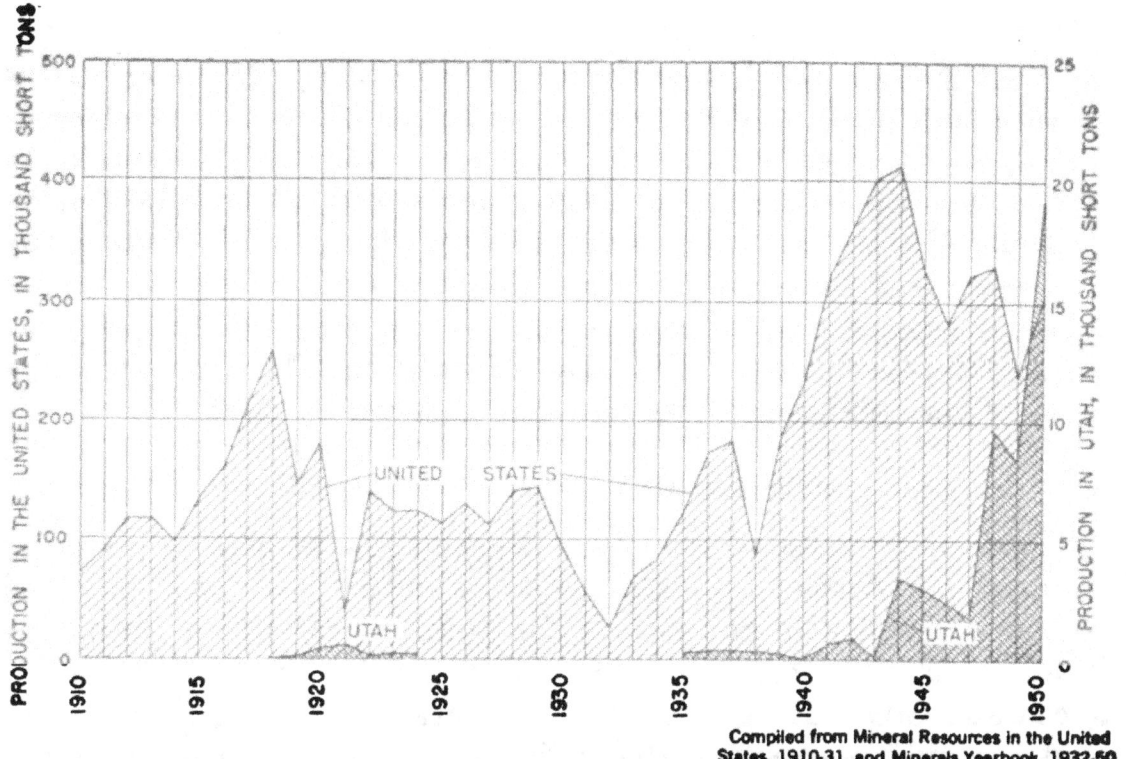

FIGURE 2.—Fluorspar production of Utah compared to national fluorspar production.

Compiled from Mineral Resources in the United States, 1910-31, and Minerals Yearbook, 1932-50

made in cooperation with the State of Utah through the University of Utah. The study of the Thomas Range district in 1950, however, was part of an investigation of uranium resources being carried out by the U. S. Geological Survey on behalf of the Atomic Energy Commission.

Most of the field work that this report is based upon was done by the writers, assisted at various times by J. O. Fisher, Elliot Gillerman, D. M. Henderson, and E. D. Washburn. The work was under three supervisors: L. R. Page, R. E. Van Alstine, and J. S. Williams.

HISTORY OF FLUORSPAR MINING IN UTAH

The first recorded mining of fluorspar in Utah was at the Silver Queen mine near Clive, Tooele County; some 20 tons was sold in 1918. In the period 1918–1924 about 900 tons of saleable fluorspar was produced from the Silver Queen mine. In the next decade no fluorspar was mined in the State. In 1935 production was reported from the Monarch and Indian Peak areas in Beaver County, and between 1935 and 1945 Beaver County was the largest producer of fluorspar in Utah. In May 1943 fluorspar was first produced from claims located in the Thomas Range in Juab County, and by September 1950 some 35,700 tons had been produced from that district. The production data are summarized in figure 2.

CLASSIFICATION OF THE DEPOSITS

Most of the fluorspar deposits of Utah are epithermal deposits filling fault fissures and interstices in fault breccia. At many of the deposits fault movement was recurrent during the period of mineralization, and locally the fluorite is brecciated. Some of the deposits in the Thomas Range district of Juab County and in the Star district of Beaver County are replacement deposits; the irregular boundaries of fragments of limestone and rhyolite porphyry enclosed in massive fluorspar suggest partial replacement at the Monarch (Staats) deposits of Beaver County. Evidences of wall-rock alteration generally have not been observed in the fluorspar districts of Utah.

MINERALOGY OF THE DEPOSITS

Fluorite.—In most of the Utah deposits the fluorite is massive; however, well-formed crystals, generally cubes, have been observed in Beaver County in deposits of the Indian Peak Range, Monarch (Staats) claims, and Star district, and in Tooele County at the Silver Queen and Silver King deposits. The grain size ranges from very coarse, as in the deposits of the Indian Peak Range, to extremely fine as in the Monarch (Staats) deposit and in the Thomas Range district, Juab County. Much of the Thomas Range fluorite is soft and powdery.

The fluorite ranges from colorless to green, blue, and dark purple in the Beaver County deposits, from white to purple in the Juab County deposits, and from almost colorless to blue and green in the Silver Queen deposit, Tooele County. At the Monarch (Staats) deposit some of the dark-purple massive fluorite, when freshly broken, releases a gas that may be free fluorine.[1]

Silica minerals.—Quartz occurs as fissure fillings, crusts, and selvages on rock fragments in the deposits of the Indian Peak Range and fills holes and cracks in brecciated fluorite. Most of the quartz is massive and milky white to clear. Prismatic and finely granular varieties are common, however, and in many localities form crusts around breccia fragments. An intergrowth of lamellar quartz and calcite with orthoclase was found at the JB mine. Chalcedony locally was formed after the fluorite and quartz were deposited in veins of the Indian Peak Range.

Quartz locally fills openings in fluorite and the dolomite wall rock in the Thomas Range district; in some deposits the silica minerals and fluorite form a fine boxwork structure. A streak of reddish opaline material is along the margin of one of the veins at the Silver Queen property, Tooele County.

---

[1] Determination by Michael Fleischer, U. S. Geological Survey.

Carbonate minerals.—Lamellar intergrowths of white calcite and quartz were observed in the Indian Peak Range; most of the calcite is the white massive type, however, and formed later than the fluorite. A dark-brown fibrous calcite is also associated with fluorite in these deposits. In the Thomas Range district dolomite commonly occurs in the matrix of the granular fluorite, but the carbonate minerals generally make up less than 5 percent of the ore. White coarsely crystalline aragonite forms veins in the dolomite next to some of the fluorspar deposits in this district.

Other carbonate minerals associated with fluorite in some of the Utah deposits are witherite and rhodochrosite. Crystals of dolomite and white witherite have been found at the Silver Queen property. At the Moscow mine in the Star district rhodochrosite occurs with the fluorite.

Uranium minerals.—The fluorspar deposits in the Thomas Range district are abnormally radioactive, but generally no uranium minerals can be seen. Carnotite is present in the fluorspar at the Eagle Rock property, the Floride property, and in the West pipe at the Fluorine Queen property. At the Bell Hill and Harrisite properties it is also visible on the dolomite wall rock but not in the fluorspar. In the Monarch (Staats) deposit of Beaver County, Wyant[2] found autunite and uranophane locally as coatings on the fluorite.

Other minerals.—Clay minerals constitute a variable, but generally the largest, part of the gangue in the Thomas Range and Monarch (Staats) fluorspar deposits. Orthoclase, probably in the form of adularia,[3] makes up about 10 percent of an intergrowth of lamellar quartz and calcite in a specimen from the JB mine in the Indian Peak Range. At various places within a 90-acre tract surrounding the Silver Queen property, barite, pyrite, chalcopyrite, and traces of gold and silver have been reported (Burchard, 1933, p. 21) in veins with fluorite.

Sulfide minerals are associated with fluorite elsewhere in Utah. Pyrite occurs in the wall rock of the fluorspar veins at the Monarch (Staats) property. In the contact-metamorphic deposits of the Star district galena, sphalerite, pyrite, and chalcopyrite accompany fluorite, rhodochrosite, magnetite, specularite, and various silicates (sericite, muscovite, garnet, epidote, tremolite, diopside, and pyroxene). A few other minerals are found in the minor occurrences of fluorite listed in table 3.

[2] Wyant, D. G., 1947, Staats fluorspar mine, Beaver County, Utah: U. S. Geol. Survey, unpublished rept., 11 p.

[3] Determination by Jewell J. Glass, U. S. Geological Survey.

DEPOSITS IN INDIAN PEAK RANGE AND MONARCH CLAIMS, BEAVER COUNTY

By W. R. Thurston

INDIAN PEAK RANGE
GENERAL GEOLOGY

The main producing area of Beaver County is in the Indian Peak Range (fig. 3) partly in the old Washington metal-mining district but mainly northeast of it. The main fluorspar producing properties are the Cougar Spar and Blue Bell mines in T. 30 S., R. 18 W., and the Utah mine and JB claim in T. 30 S., R. 17 W. Scattered exposures of fluorite-bearing material are found over an area of about 20 square miles.

The fluorspar deposits are in an area of volcanic flows and tuffs that have been intruded by a quartz diorite stock. One small area of quartzite was noted, but no other sedimentary rocks were found. The flows range in composition from rhyolite to andesite. Microscopic examination of the extrusive rocks in the Cougar Spar-Blue Bell areas (pl. 1) shows that the majority are latites. These latite flows have dacitic, andesitic, and trachytic facies, show considerable alteration, and contain many inclusions. The inclusions have the same range in composition as the flows and represent both preexisting material picked up by the moving lava and chilled parts broken off the main flow. In some localities flows are separated by lenses of pyroclastic material. Many perlitic bands separate the rhyolite flows. The tuffs are cemented volcanic ash containing abundant quartz grains. They are extensively devitrified, and in some exposures are difficult to distinguish from the rhyolite.

The intrusive rock is a quartz diorite that contains a small amount of orthoclase. The quartz diorite is epidotized near the fault zone at the Cougar Spar mine. Discontinuous vertical rhyolite dikes, striking in a northeasterly direction and ranging in width from 4 to 20 feet, occur in the quartz diorite. The rhyolite is generally fine grained and banded parallel to the walls.

A small dacite flow, near latite in composition, rests on eroded quartz diorite on the ridge southwest of the Cougar Spar mine. Most of the igneous rocks of western Utah and eastern Nevada generally are regarded as being of Tertiary age, but this dacite flow may be of Quaternary age.

The rocks of the area are broken by fault and shear zones. Hydrothermal solutions, probably of the epithermal type, altered the country rock and deposited quartz, calcite, and fluorite in the zones of dislocation. The mineralized zones are as wide as 250 feet but do not have well-defined walls; the veins are biggest and most common near the

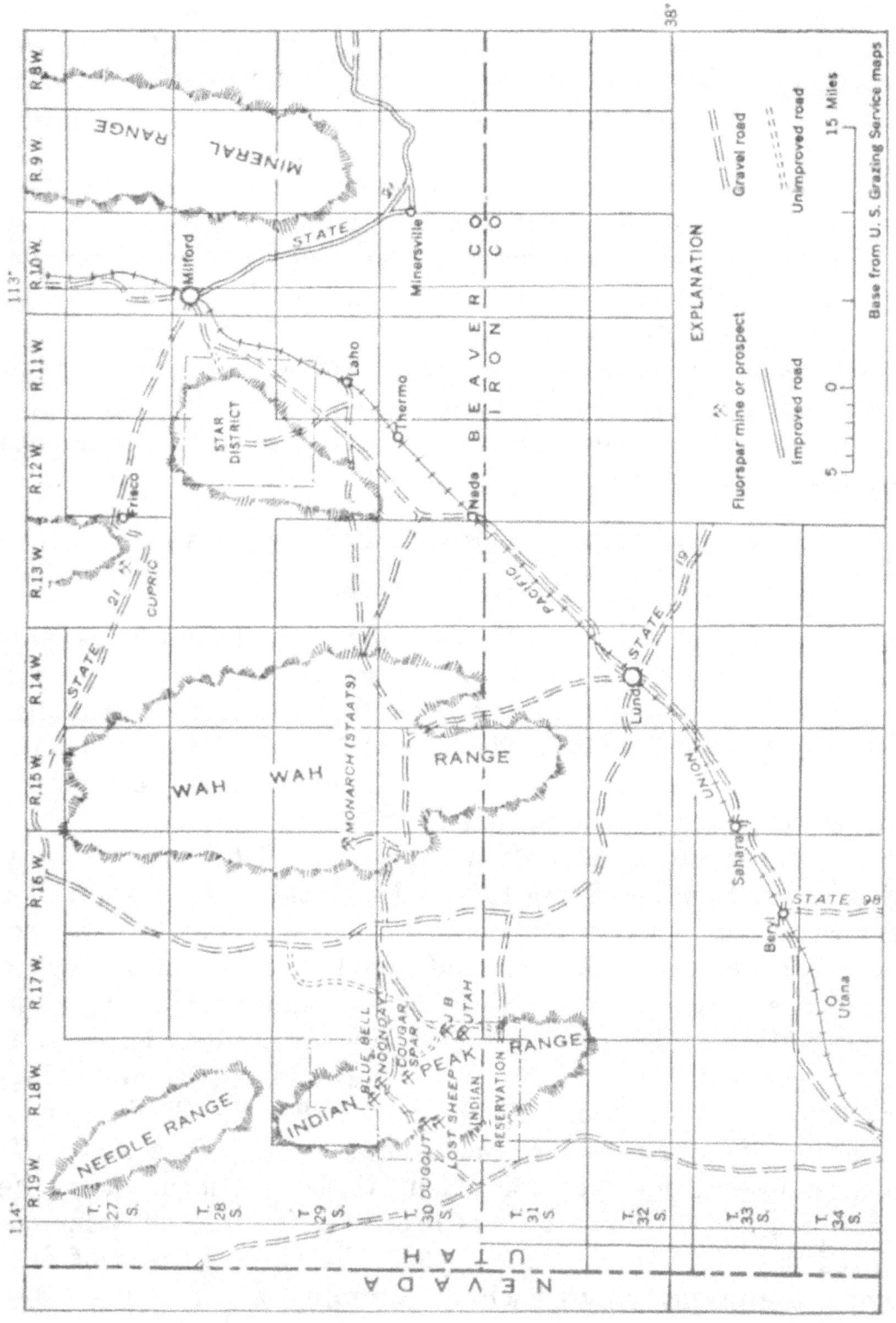

FIGURE 3.—Index map showing fluorspar localities in Beaver County, Utah.

center of these zones and grade outward through progressively less brecciated and less mineralized rock to almost unaltered country rock. Movement along the faults was recurrent throughout mineralization, and deposition did not proceed uniformly, as indicated by many sharp structural truncations and mineral repetitions in the veins. The vein minerals are locally crushed, and the veins contain many vugs and openings. The fluorspar occurs most abundantly in the highly deformed parts of the fault zones, although small amounts of fluorite are present with other vein minerals throughout the breccia. The fluorspar bodies vary from lenses and veins of almost pure fluorite to coarsely crystalline aggregates of quartz, calcite, and fluorite containing rock fragments. Individual veins and lenses of fluorspar pinch and swell within short distances and lose identity by grading laterally and vertically into other vein material.

The general order of mineralization, as revealed in mine workings, appears to have been: fine-grained quartz, white lamellar calcite, granular quartz, prismatic quartz, fluorite, massive quartz, chalcedony, massive white calcite, and dark-brown calcite.

DESCRIPTION OF INDIVIDUAL DEPOSITS

COUGAR SPAR MINE

The Cougar Spar mine is on the east side of the Indian Peak Range in SE¼ sec. 10, T. 30 S., R. 18 W., at an average altitude of 7,800 feet (fig. 3). The mine may be reached by graded roads from either Milford in Beaver County, a distance of 57 miles, or from Lund in Iron County, a distance of 35 miles. Lund is the nearest shipping point on the Union Pacific Railroad.

W. C. Dalton and S. M. Dalton of Parowan, Utah, staked a claim at the Cougar Spar deposit in 1937 and sank the inclined shaft of the upper workings. In 1941 Hubert C. Eyre of Minersville, Utah, relocated the claim and staked an additional claim on each side: the three claims were named the Spar, Spar 1, and Spar 2. E. A. McKenzie of Minersville and Fred Morrman and B. M. Anderson of Los Angeles leased the claims from Eyre in 1941 and 1942 and shipped about 600 tons of fluorspar. In the fall of 1942 the Tintic Standard Mining Co. of Salt Lake City acquired the property from Eyre. The three claims were surveyed for patent, three additional claims were located, and two 40-acre parcels of ground adjoining the claims were acquired (fig. 4). In the fall of 1943, the U. S. Bureau of Mines explored the property by trenching, sampling, and drifting (Marsh and Everett, 1945; Everett and Wilson, 1951), and the Tintic Standard Mining Co. built a 150-ton-a-day jig mill and started operations.

The following Geological Survey men have worked on the geology of the Cougar Spar area: D. C. Cox and J. O. Fisher mapped the

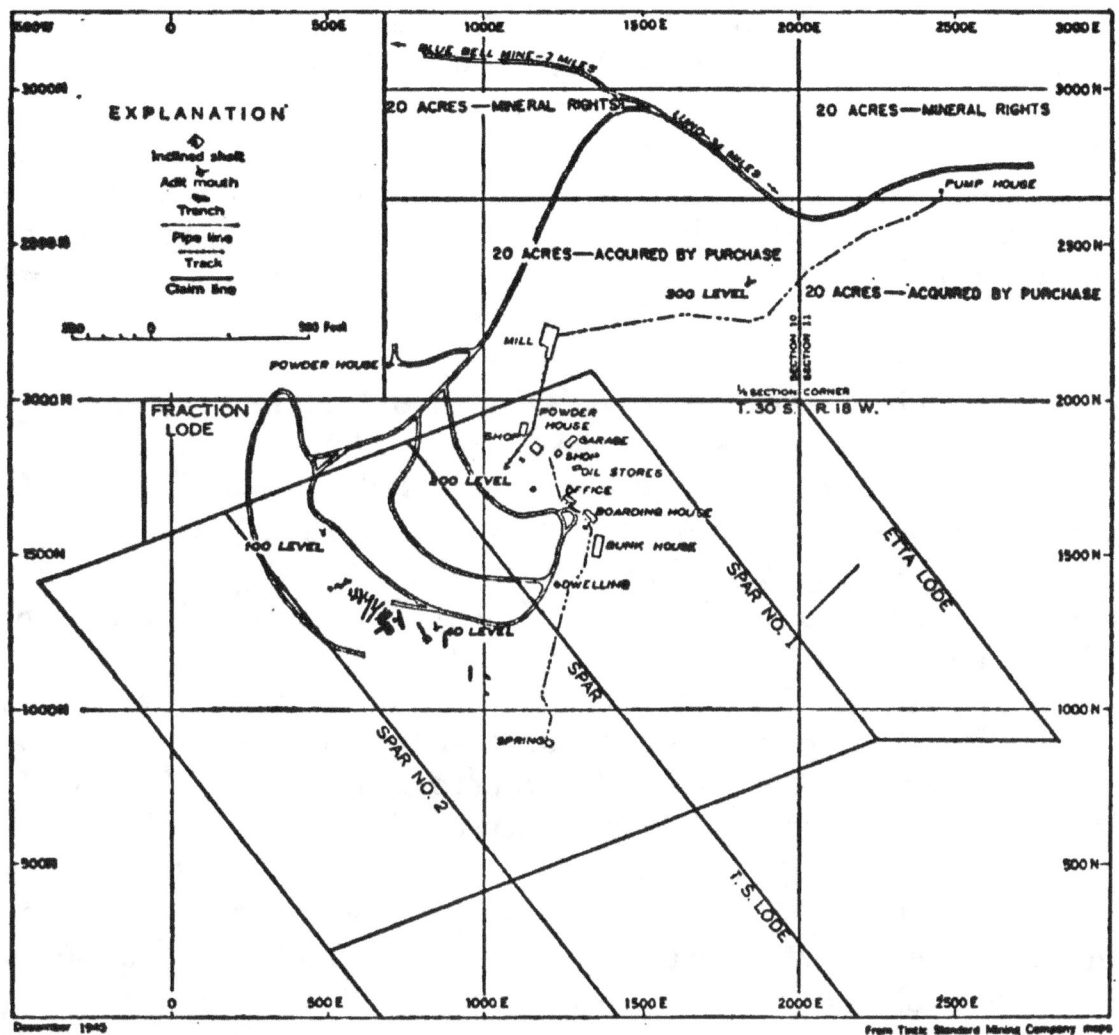

FIGURE 4.—Map showing claims and surface plant, Cougar Spar mine, Beaver County, Utah.

trenches and other workings of the Bureau of Mines in April and June 1943; A. E. Granger continued this work until July 1944; from July 1944 through December 1945 the writer recorded the developments at the Cougar Spar mine. From August through December 1945 the topography and geology of the area (pl. 1) were mapped by the writer, assisted part of the time by D. M. Henderson and part of the time by Elliot Gillerman. Contacts are poorly exposed, and most of the mapping had to be done on the basis of float distribution. Mapping was further hampered by snow.

The fluorspar deposit at the Cougar Spar mine is in a fault zone near the contact between an intruded quartz diorite and rock of a volcanic series (pl. 1). The fault zone strikes N. 15° W. and is traceable for at least a mile both north and south of the mine. South of the mine the quartz diorite is in fault contact with various members, chiefly pyroclastic, of the volcanic series; north of the mine for 2,000 feet it is in contact with extrusive rocks. Farther north the fault zone continues within various latitic pyroclastic rocks and flows. The surface trace indicates that the fault zone has a steep but variable dip to the east.   The trace is marked by a belt of quartz veins and quartz-filled breccia; in places lamellar calcite may be found, but detrital fluorite is rare.   At the Cougar Spar mine the surface of the fault zone was marked by an especially broad belt of overburden containing vein quartz and quartz-filled breccia.   A ledge of quartz-cemented breccia from 2 to 4 feet wide rises as much as 5 feet above the ground for a distance of 200 feet along the west side of the fault zone.   Detrital fluorspar is not common even above the known deposit; so, the possibility of additional fluorspar bodies north and south of the workings cannot be eliminated because of the scarcity of fluorspar float.

The Cougar Spar deposit was first explored by an inclined shaft, surface trenches, several crosscuts, and adits at the 40- and 100-foot levels (pl. 2 and fig. 4).   These workings yielded about 600 tons of fluorspar.   The Tintic Standard Mining Co. developed the mine with an adit at the 200-foot level, which leads directly to the jig mill (pl. 2 and fig. 4).   An adit at the 300-foot level was started late in 1945.   The workings on the 200-foot level consist of an adit 700 feet long, and almost 900 feet of drifts and crosscuts (pl. 2).   The workings extend about 500 feet along the strike of the fault zone.

In the mine a fault zone more than 200 feet wide on the 200-foot level strikes about N. 30° W. and dips steeply east.   A white tuff forming the north wall at the west end of the mine is silicified and sliced into several blocks by small branch faults (pl. 2).   The blocks of tuff near the quartz diorite on the southwest are altered to a gray soft rock. The fault zone appears to contain more fragments and blocks of the quartz diorite than of the tuff; however, the finer portions of the gouge may have been derived from the tuff, because the tuff shows a greater tendency to alter and weather to a soft material than does the quartz diorite.   The fissures and breccia were filled with quartz, calcite, and fluorite.   Postfluoritization movement did not make any major changes in the fault as a whole but truncated the fissure fillings in many places, offsetting the fluorspar-bearing zones from 5 to 30 feet.   The offsets are greater and more distinct near the surface than they are at depth.

BLUE BELL MINE

The Blue Bell mine is in SE¼ sec. 4, T. 30 S., R. 18 W., about 6,000 feet northwest of the Cougar Spar mine and at an altitude of 8,500 feet (pl. 1 and fig. 3).   The Blue Bell property includes 24 unpatented claims (fig. 5), which were located in the spring of 1937 by H. J. Holt of Milford, Utah.   Holt was the operator and part owner of the Blue Bell mine.

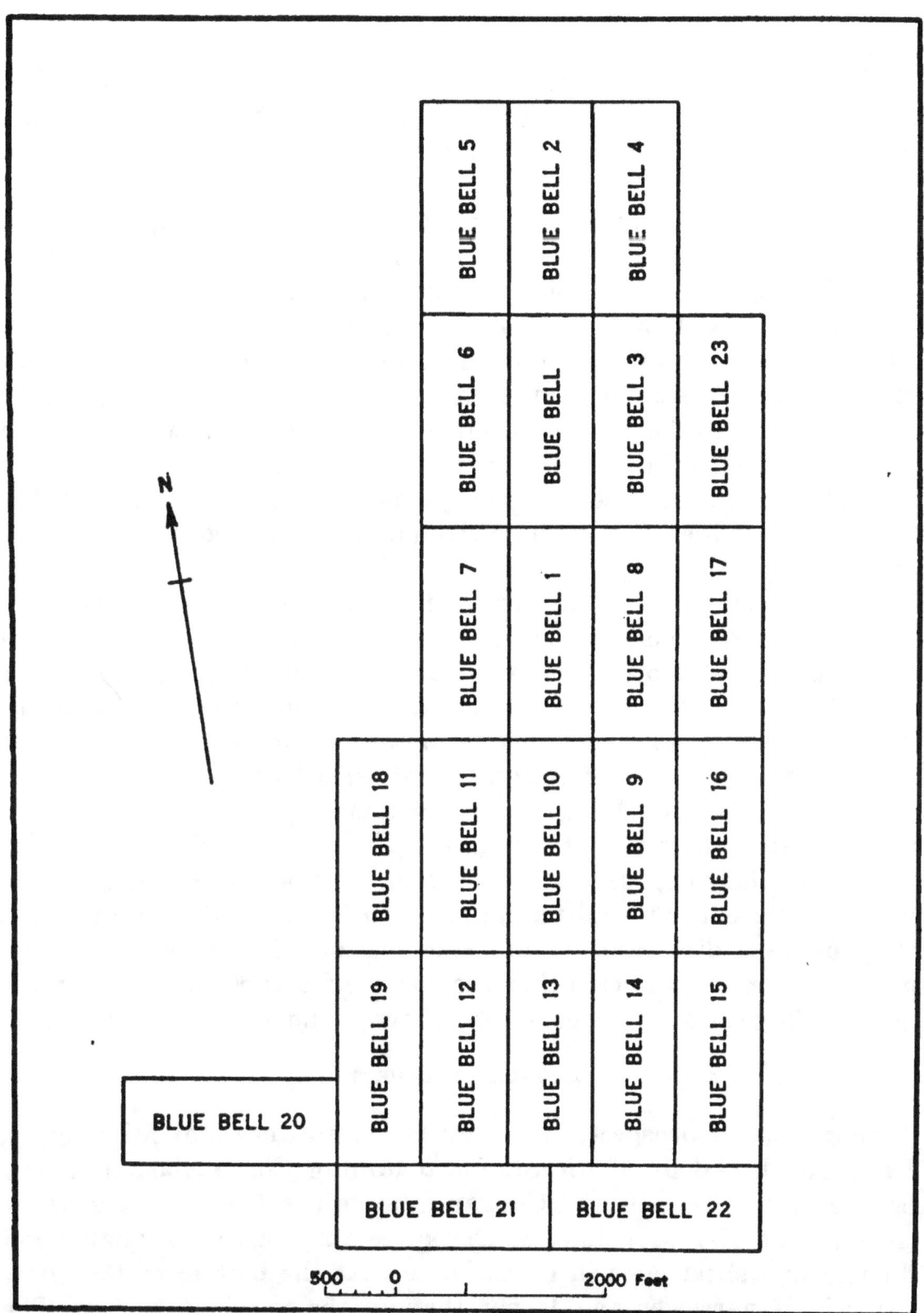

FIGURE 5.—Plat of Blue Bell claims, Beaver County, Utah.

In April 1943, D. C. Cox and J. O. Fisher visited the Blue Bell property as part of the Geological Survey's program for the investigation of strategic minerals. In August 1944 the writer, assisted by E. D. Washburn, mapped the topography and geology in the immediate vicinity of the mine workings.

The workings on the property consist of 2 shafts, several pits and trenches, a series of bulldozer cuts, and 1 adit (pl. 3). The main shaft is vertical and 50 feet deep; at the bottom of the shaft are 40 feet of irregular drifts. This shaft was not in use because of the inflow of water. The other shaft is inclined 65° NW. for 30 feet; a 100-foot drift extends northward from it. On the surface near the shafts an area 180 feet long and roughly 30 feet wide was levelled and partly stripped by a bulldozer. All of the surface and underground work has been done in an area about 330 feet long and 30 feet wide, in search of minable widths of high-grade ore. In the workings the relations of the fluorite to the other vein minerals are not well exposed. The quartz, calcite, and fluorite occur as vertical lenses striking N. 35° W. in a shear zone in greenish-gray porphyritic latite. Most of the workings have been driven along the fluorspar-rich parts of the shear zone.

Deeper workings were driven by the U. S. Bureau of Mines. A 460-foot adit cut the fluorspar-bearing zone 400 feet from the portal and at a vertical depth of 140 feet below the collar of the main shaft (pl. 3), and drifts were driven from this adit. At this depth the fluorspar-bearing zone is 12 feet wide and forms part of a sheared and mineralized zone 40 feet wide. Only small stringers of fluorspar were found in the workings at this level, and appear to be fewer and narrower than those near the surface. The east side of the fault zone is vertical where exposed in the adit and was not cut elsewhere. The west side was cut in the adit and in the south drift and crosscut. The steep opposed dips recorded and the relation of surface and underground exposures indicate that the shear zone is essentially vertical despite the varied attitudes of the internal and marginal structures.

NOONDAY PROSPECT

The Noonday prospect, owned by H. A. McKenzie of Minersville, Utah, and leased by Tintic Standard Mining Co., is about halfway between the Blue Bell and Cougar Spar mines (pl. 1). A shallow trench and a 30-foot inclined shaft expose a 2-foot vein of quartz and fluorite in a shear zone in green latite. At the bottom of the shaft the vein is narrower and leaner than in the trench (fig. 6). The vein strikes N. 10° W. and dips 45° W. It was not possible to locate this fluorspar vein either north or south of the prospect workings, and similar veins were not found nearby.

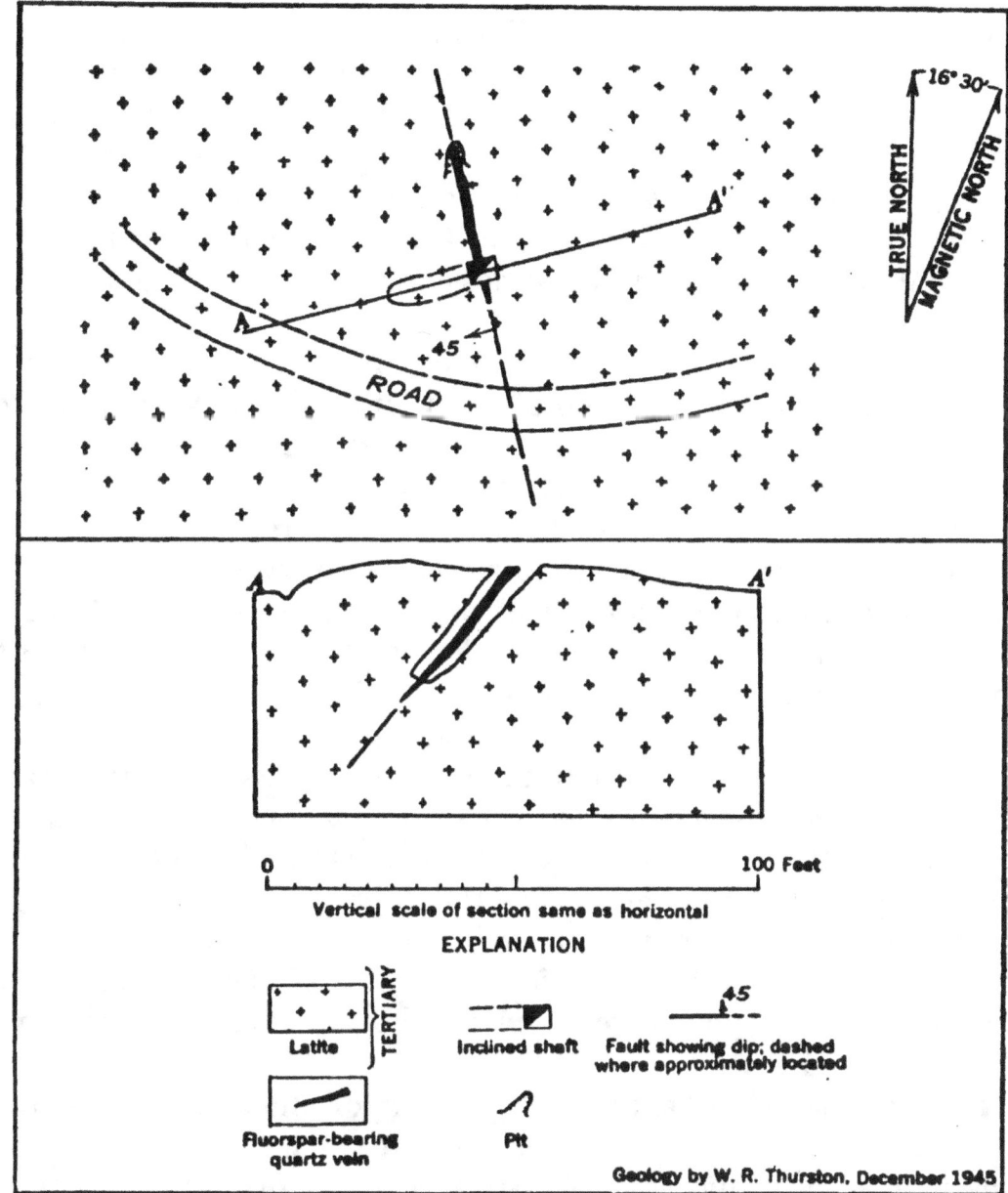

FIGURE 6.—Sketch map of Noonday prospect, Beaver County, Utah.

JB PROSPECT

The JB prospect is in sec. 30, T. 30 S., R. 17 W., about 3½ miles southeast of the Cougar Spar mine (fig. 3). The prospect may be reached by an unimproved road near the Cougar Spar property or by a branch of the main access road to the Cougar Spar mine.

The JB claim was located in July 1942 by Bart W. Mortensen, John Davenport, Gene Bullock, and Vera Mortensen, all of Parowan, Utah. It has been leased for an indefinite term by Theo. E. Stevens of Milford, Utah. Surface trenching by the lessee between July 1942 and June 1943 yielded about 30 tons of high-grade fluorspar. In June 1943, D. C. Cox and J. O. Fisher mapped the trenches on the JB claim and traced the vein-zone from them to the Utah mine. The Bureau of Mines started underground exploration at the JB prospect in October 1943 (Everett and Wilson, 1950).   A. E. Granger of the Geological Survey made frequent visits to the JB prospect while the Bureau of Mines was doing the exploratory work.   In July 1944 the writer, assisted by E. D. Washburn, mapped the topography and geology of the JB area and made detailed geologic maps of the surface and underground workings (pl. 4).   The writer kept these maps up to date until the Bureau of Mines completed the exploratory work in October 1944.

The workings at the JB prospect consist of several trenches over an area about 250 feet long and 200 feet wide; the underground workings consist of a 106-foot shaft, inclined 48° SE., with about 550 feet of drifts and crosscuts at a vertical depth of 63 feet below the collar of the shaft and from 70 to 110 feet below the trenched area.

The JB prospect is in a broad fault zone between a red dacite on the northwest and a green andesite on the southeast.   The fault zone consists of a fissured and brecciated belt at least 250 feet wide that trends N. 60° E. for an undetermined distance.   Within this belt zones of veins ranging in trend from N. 10° W. to N. 50° E. and dipping steeply to the north and northwest, consist of quartz, calcite, and fluorite.   These zones are not continuous or parallel, and pinch and swell within short distances.   The waste rock between zones of veins is brecciated and veined, but the fluorspar content is negligible. Massive quartz veins are common in the area, and one vein forms a ridge as much as 10 feet high for a distance of 200 feet northeast of the shaft.   This quartz vein marks the easternmost limit of the fluorspar-bearing zones.

The trenches show a high proportion of fluorite relative to the amount of quartz and brecciated andesite.   The underground workings expose fewer fluorspar-rich zones, and the proportion of country rock and gangue material is much higher.

UTAH MINE

The Utah mine is a mile southwest of the JB claim and may be reached by way of the same entrance roads (fig. 3).   The Utah mine is on the Crystal claims, located in 1936 by the Utah Fluorspar Corp., of which Bart W. Mortensen, Parowan, Utah, is president.   The claims are under lease to Theo. E. Stevens and R. H. Alsop of Milford, Utah.   In April and June 1943 D. C. Cox, J. O. Fisher, and A. E. Granger of the Geological Survey visited the Utah mine.   The description of the workings (fig. 7), now inaccessible, is taken from their notes.

An almost vertical fault zone, striking N. 45° E., separates andesite porphyry on the southeast from rhyolite porphyry on the northwest. The fault zone is about 80 feet wide.   The greater part of this width consists of blocks and masses of altered andesite with veins and fillings

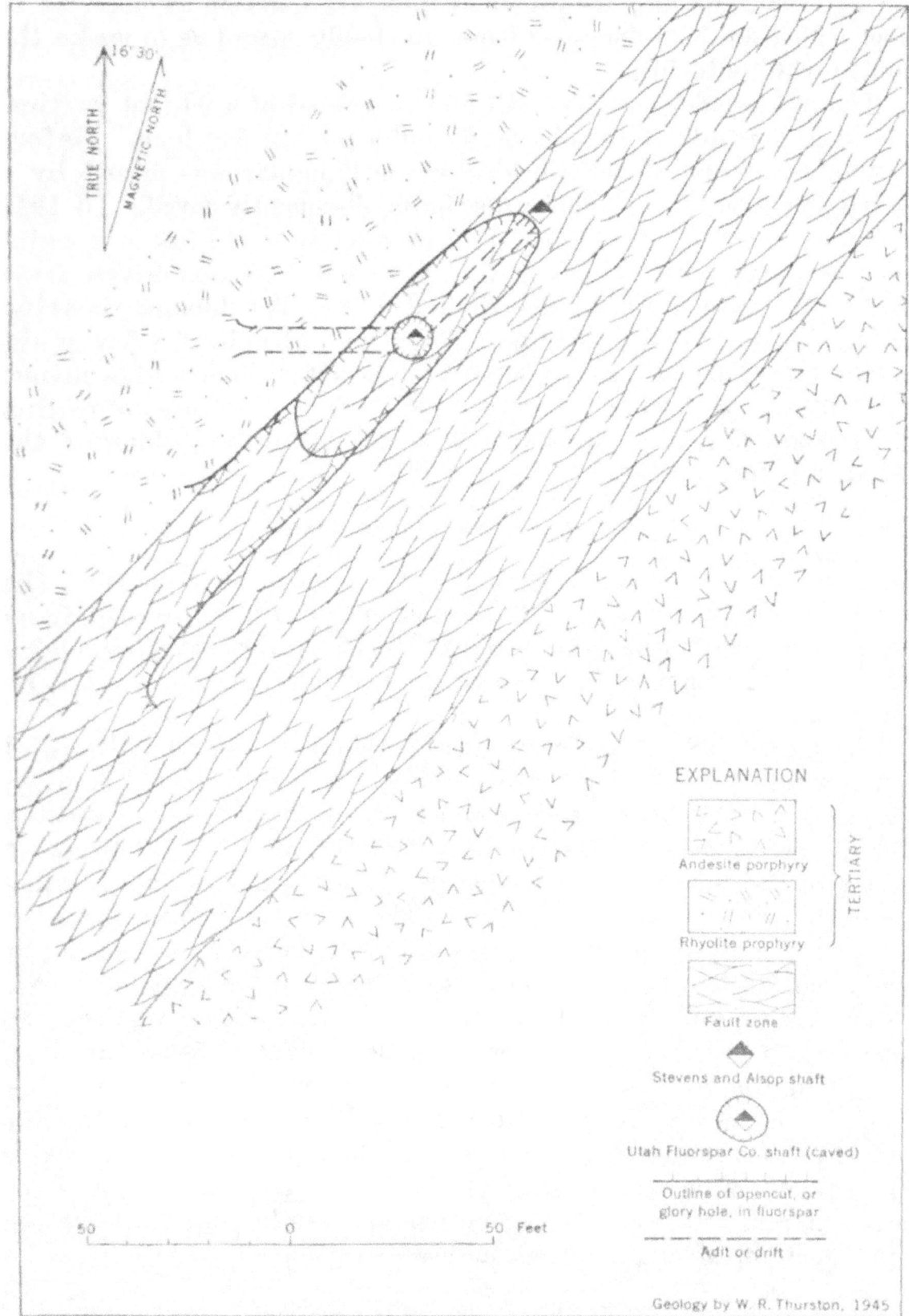

FIGURE 7.—Geologic sketch map of the Utah mine, Beaver County, Utah.

of quartz and calcite. Quartz and calcite are traceable for about 500 feet in both directions along the strike. Along the northwest margin of the fault zone fluorspar is present as seams and lenses as wide as 2 feet. The fluorspar-bearing part of the fault zone is not more than

20 feet wide, and near the center of this part a section as much as 12 feet wide contained fluorspar lenses so closely spaced as to make the entire width minable.

The old workings at the Utah mine consisted of a 90-foot vertical shaft, a short adit to the old shaft, and a pit 110 feet long. Before 1940, 12 carloads of metallurgical-grade fluorspar was mined by a glory hole from the old shaft; the shaft subsequently caved. In 1942 Stevens and Alsop sank a new shaft to a depth of 110 feet at a point 45 feet northeast of the old shaft. A 40-foot drift was driven from the 90-foot level to the old glory hole (fig. 7). The fluorspar-bearing zone at the new shaft was reported by Cox to be about 8 feet wide, about half of which was fluorspar in 1-foot lenses that could be mined selectively. According to Stevens, 5½ carloads of ore averaging more than 80 percent of CaF_2 was mined during the sinking of the shaft and the drifting.

LOST SHEEP AND DUGOUT CLAIMS

The Lost Sheep and Dugout claims are on the west slope of the Indian Peak Range (fig. 3). They are about 5 miles by road from the Cougar Spar mine. E. A. McKenzie leased the Lost Sheep claim from J. H. Haslam, and the Dugout claims from C. L. Jones and J. H. Haslam.

At the Lost Sheep claim a wide vein zone containing calcite and quartz cuts the volcanic rock of the area. A few weathered outcrops suggest that the country rock is andesitic. A 25-foot vein of massive calcite occurs near the eastern margin of the vein zone. Veinlets of fluorspar are exposed in two outcrops, one 200 feet west and the other 500 feet west of the calcite mass. The fluorspar veinlets occur in quartz and range in width from 1 to 6 inches. No economically minable width of fluorspar is exposed. The vein zone extends northward a few hundred feet to a gully, as shown by float of quartz and calcite. Beyond the gully there is little evidence of the vein zone.

At the Dugout claims a few veinlets of fluorspar occur in a quartz vein. The veinlets average 3 inches in width and are limited to zones less than 4 feet wide. One exposure of fluorspar-bearing material is in a pit near the entrance road; a shallow shaft 15 feet southeast of the pit revealed no fluorspar. Fluorspar veinlets are also exposed in an outcrop about 1,000 feet southeast.

MONARCH (STAATS) CLAIMS

The Monarch (Staats) claims are in the Wah Wah Mountains, 47 miles by road southwest of Milford, Utah, and about 35 miles by road northwest of Lund, Utah (fig. 3). The mine is in T. 29 S., R. 16 W., in what is sometimes called the Pine Grove mining district. The claims are owned by Fred Staats of Salt Lake City. Fluorspar was first discovered in the area about 1935, and some ore was mined there every year through 1945. The total production to January 1946 is about 3,500 tons. The claims have been in several hands and called by various names: the Monarch mine, the Skougard mine, the Roberts and Skougard mine, and the Roberts and Staats mine. Recently the area has been referred to solely by the name of the present operator and is called the Staats area. In April 1943 Cox and Fisher visited the Staats mine and mapped the relation of the pits and trenches to the upper shaft. In March 1945 the writer examined the underground workings and mapped part of them.

The fluorspar deposits occur along the faulted contact of a Tertiary rhyolite porphyry and a Cambrian limestone. The wall rocks were altered only slightly, but irregular boundaries of some fragments of country rock enclosed in massive fluorspar suggest partial replacement. The silica content of the ore is low and is probably derived from inclusions of rhyolite porphyry; quartz veins are not present in the ore. Some $CaCO_3$ is present, owing to included limestone fragments. The SiO_2 and $CaCO_3$ content did not exceed 5 percent, respectively, and averaged about 2 percent of each. Autunite and uranophane were found by Wyant[4] as local coatings on the fluorite. Two grab samples of material from some of the pits northeast of the main shaft contained 50 and 68 percent of CaF_2. Assays of carload shipments of fluorspar that had been selectively mined and sorted ranged from 80 to 91 percent of CaF_2 and averaged in excess of 85 percent of CaF_2.

The fluorspar occurs in lenticular shoots within larger podlike ore zones. The waste between shoots is composed of brecciated limestone and rhyolite porphyry. Apparently the depth of each shoot is greater than the length: the shoots range from 2 to 6 feet wide and from 5 to 10 feet long, but are reported to extend more than 25 feet in depth. The shoots are oriented roughly parallel to the irregular contact zone of the limestone and rhyolite porphyry. The contact zone is intricately faulted, with many variations in strike and dip and sharp undulations in the trace.

Two separate areas of fluorspar mineralization on the Staats property have been explored. Most of the prospecting centers around the main deposit at the head of the valley (fig. 8) but a small deposit has been opened about half a mile southeast (fig. 9).

The principal workings at the main deposit (the Staats mine) are an 85-foot shaft, an opencut, and an adit. This deposit was first explored by an opencut; then a shaft was sunk about the middle of the opencut, and stopes were driven at various levels from the shaft. At

---

4. Op. cit. on p. 5; p. 8.

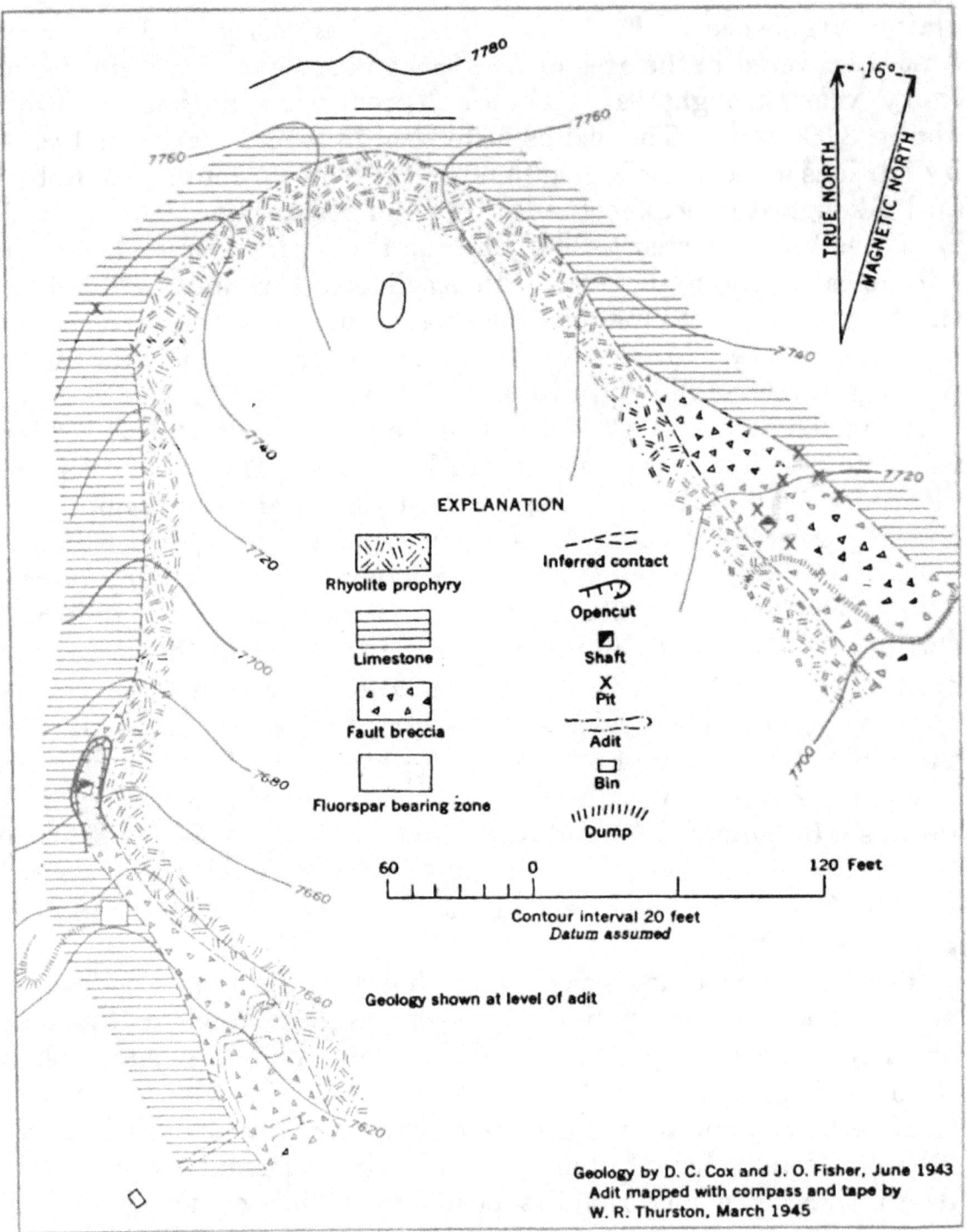

EXPLANATION

Rhyolite prophyry

Limestone

Fault breccia

Fluorspar bearing zone

Inferred contact

Opencut

Shaft

Pit

Adit

Bin

Dump

60   0   120 Feet

Contour interval 20 feet
Datum assumed

Geology shown at level of adit

Geology by D. C. Cox and J. O. Fisher, June 1943
Adit mapped with compass and tape by
W. R. Thurston, March 1945

FIGURE 8.—Sketch map of geology in vicinity of main shaft, Staats mine, Beaver County,
Utah.

the 85-foot level a drift extends 65 feet southward. About 400 feet
southeast of the shaft an adit exposed several pockets of fluorspar
along a faulted segment of the contact between the rhyolite porphyry
and limestone (fig. 8). Because of caving ground, only a small part
of the shaft-workings was accessible for examination. Apparently
enough fluorspar remains in the vicinity of the shaft to be worth
recovering. Within 600 feet to the north, northeast, and southeast,
several pits and trenches show fluorspar in the contact zone.

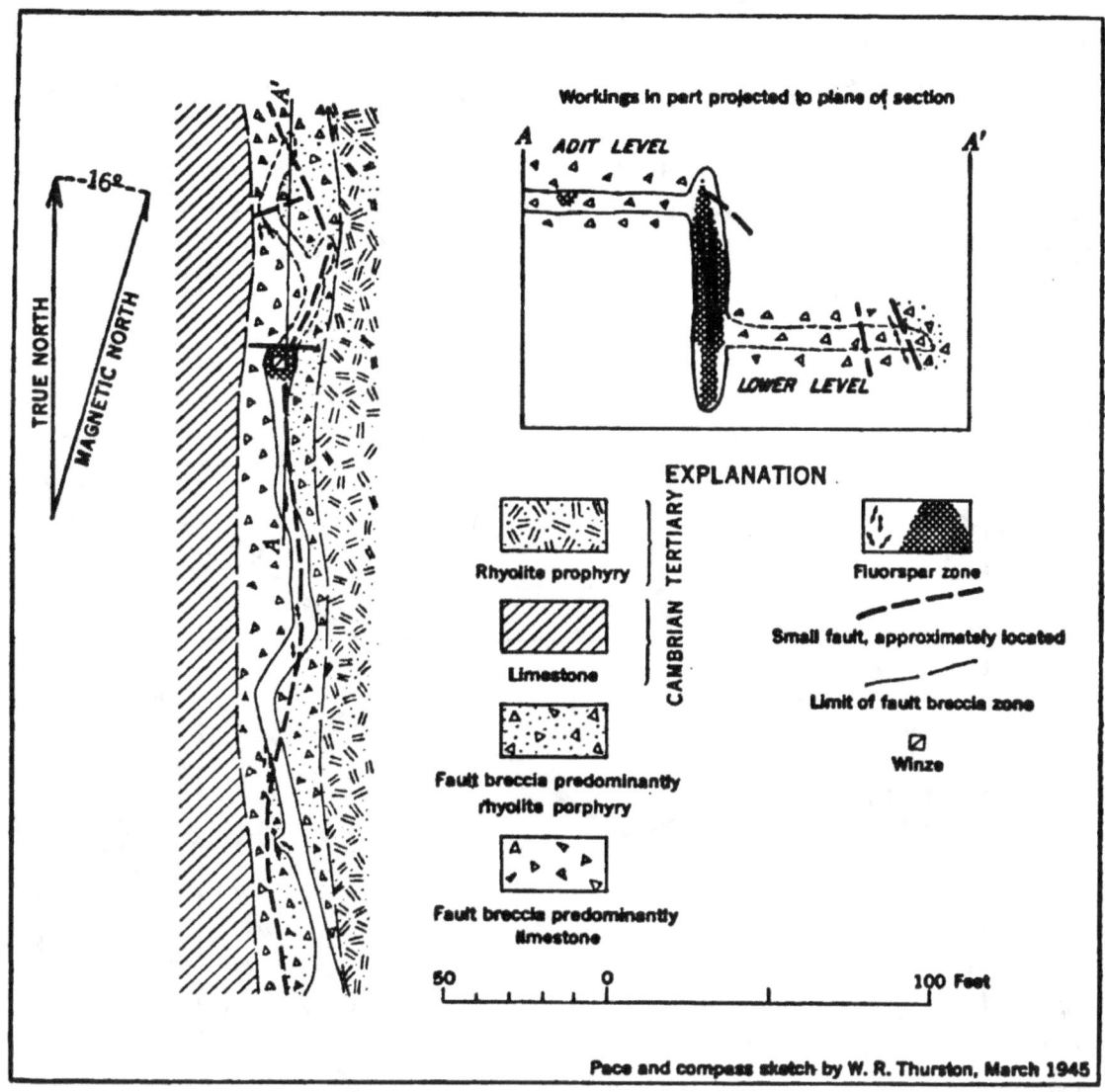

FIGURE 9.—Geologic sketch map of winze operation, Staats mine, Beaver County, Utah.

An adit about half a mile to the southeast leads to a winze operation from which 200 tons of ore was mined in 1944 (fig. 9). The ore in the winze formed a shoot 10 feet long, 6 feet wide, and about 55 feet deep.

DEPOSITS IN STAR DISTRICT, BEAVER COUNTY

By D. C. Cox

The Star district is about 8 miles west of Milford in Beaver County (fig. 3). It was one of the early copper-lead-zinc-silver mining camps of Utah and was most active during the 1870's. The district has been described by B. S. Butler (1913, p. 96, 114–136, 194–206, and pl. 1).

Fluorspar occurs in the Star district as a gangue mineral in some of the metal ores, and as seams and veins genetically related to these metal deposits but located outside them. The fluorspar in the ores of copper, lead, zinc, and silver was not recovered. The calcium fluoride content and the tonnages of ores mined at the Moscow mine in 1945 were too low to warrant separate installation of mill facilities to recover byproduct fluorspar. Fluorspar from the outlying small veins and seams is easier to concentrate and, therefore, has been mined from pits, trenches, and a few shallow shafts.

In June 1945 D. C. Cox and J. O. Fisher, accompanied by Theo. E. Stevens, examined parts of the Star district. The fluorspar-bearing areas in the district were under lease to the Western Fluorite Co., of which Mr. Stevens is president. The properties examined, and others reported to contain fluorite, are listed in table 1 and shown in figure 10.

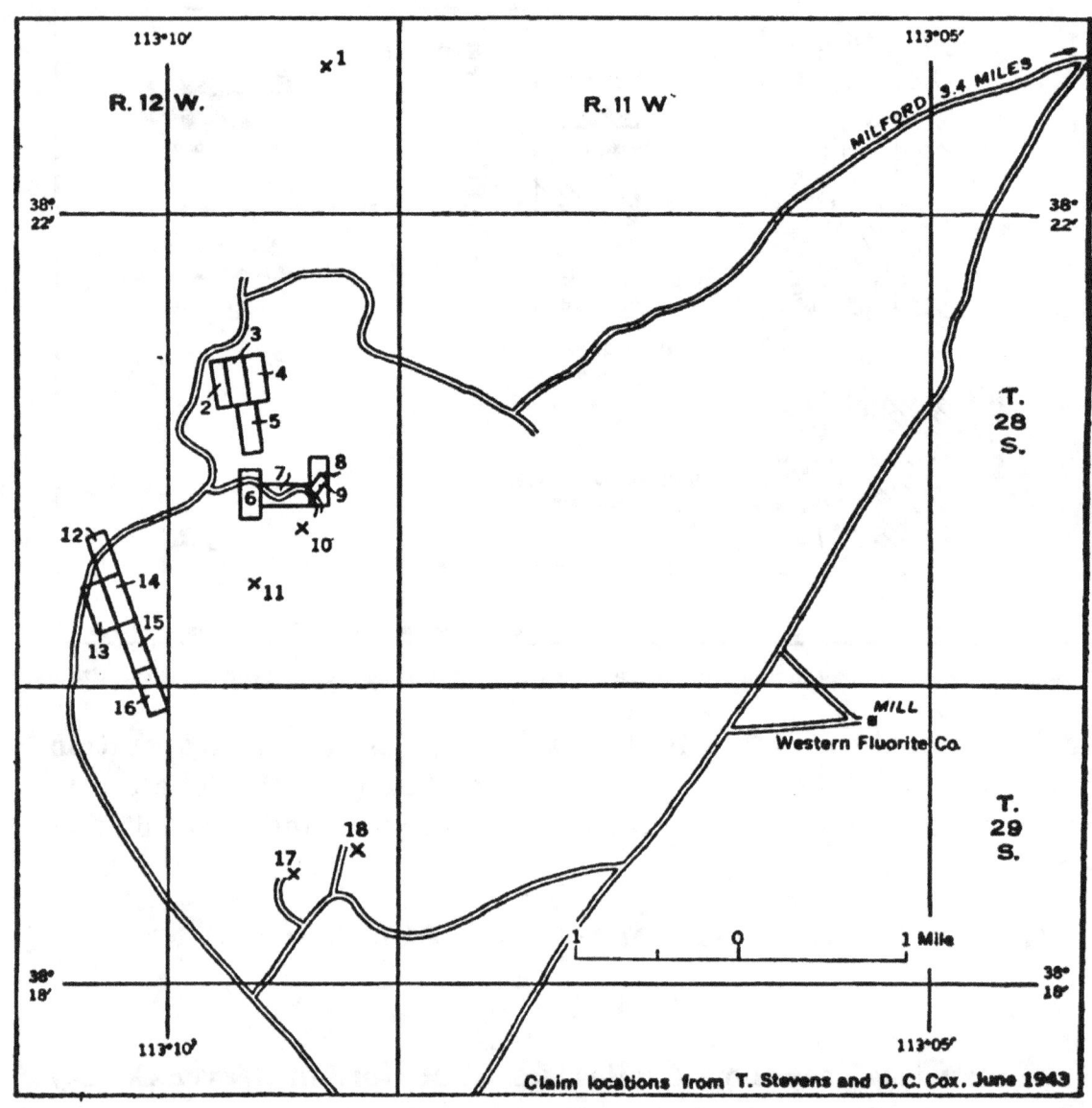

EXPLANATION

FLUORSPAR LOCALITIES

1. Wild Bill
2. Brown Thrush
3. Lucky Boy No. 1
4. Fluorine Ledge
5. Lucky Boy No. 2
6. Monte Cristo
7. Manassa
8. Local
9. Burning Moscow } Moscow
10. Hub
11. Lady Bryan
12. Virginia No. 1
13. North Virginia Extension
14. Virginia No. 2
15. Virginia No. 3
16. Virginia No. 4
17. Quartzite
18. Cabin

LOCATIONS OF ROADS IN NORTHWEST PART ARE APPROXIMATE

FIGURE 10.—Index map showing fluorspar localities in the Star district, Beaver County, Utah.

TABLE 1.—*Fluorspar localities in the Star district, Beaver County, Utah*

Claim or mine	Owner and address as of June 1943
Brown Thrush claim	Theo. E. Stevens, Milford, Utah.
Cabin claim	Do.
Fluorine Ledge claim	James Beaudino, Cedar City, Utah, and Lawrence Dotson, Minersville, Utah.
Hub mine	Unknown.
Lady Bryan mine	Do.
Lucky Boy No. 1	Do.
Lucky Boy No. 2 claim	Theo. E. Stevens, Milford, Utah.
Manassa claim	R. H. Alsop, Theo. E. Stevens, and Adam Patterson, Milford, Utah.
Monte Cristo mine	Silver Cedar Mining Co.
Moscow mine	Silver Mines Co.
North Virginia Extension claim	Theo. E. Stevens, Milford, Utah.
Quartzite claim	Ralph Myers and Lawrence Dotson, Minersville, Utah.
Virginia claims	Thomas Gillens, Minersville, Utah.
Wild Bill mine	Unknown.

The fluorite occurs as fissure fillings in limestone on the Fluorine Ledge, Lucky Boy No. 2, and Cabin claims, in quartzite on the Quartzite claim, and in quartz monzonite on the Virginia claims. On the Hub, Lady Bryan, Moscow, Wild Bill, Monte Cristo, and Manassa claims the fluorite is associated with sulfide ores and contact-metamorphic minerals. At the Wild Bill, Moscow, Lady Bryan, Monte Cristo, and Manassa mines the introduced minerals fill fissures in limestone and replace limestone near but not at the contact with quartz monzonite. Similar deposits are found at the Hub mine, but most of the deposits are confined to the contact zones.

DESCRIPTION OF INDIVIDUAL DEPOSITS

BROWN THRUSH CLAIM

On the Brown Thrush claim (fig. 10), fluospar occurs with quartz in a zone of veins in limestone, probably the Red Warrior limestone—Silurian (?) and Devonian (?). The zone, which is 15 to 20 feet wide, strikes N. 20° W., and was traced for about 100 feet. Coarsely crystalline fluorite occurs in small patches in the zone; where most abundant, it constitutes a maximum of 20 percent of the vein material. A fissure trending N. 60° W. and cutting the zone of veins contains oxidized copper, lead, and zinc minerals, according to Stevens. The dump from the 40-foot shaft on the fissure vein contains sulfide minerals. A little medium-grained fluorspar forms gangue in the oxidized and sulfide ores.

CABIN CLAIM

On the Cabin claim (fig. 10) a 30-foot inclined shaft exposes several drusy quartz veins, 1 foot thick, containing coarsely crystalline fluorite in seams less than 5 inches wide. The veins lie along bedding planes in limestone, probably the Elephant limestone (Carboniferous) which was mapped by Butler to a point within 1½ miles north of the Cabin claim. The limestone strikes N. 5° W. and dips 80° E. Southward the veins end at a fault striking N. 75° E. and dipping 85° N. A little fluorspar occurs along this fault.

FLUORINE LEDGE AND LUCKY BOY NO. 2 CLAIMS

Fluorite and quartz occur on the Fluorine Ledge and Lucky Boy No. 2 claims (fig. 10) in irregular fissure veins and along bedding planes of a limestone mapped as Red Warrior limestone by Butler. The limestone beds strike from N. 20° E. to N. 20° W. They dip eastward at angles of 20° to 30° in most of the region, but at the north end of the Fluorine Ledge claim they apparently dip steeply west. The fluorite is green to colorless and coarsely crystalline. At some places it has distinct cube faces. Most of the fluorspar was produced from a pit about 15 feet long and 4 feet deep near the south end of the Fluorine Ledge claim. According to Stevens, the ore came from a vein that was 5 feet wide at the surface and 18 inches wide at the bottom of the pit. It probably followed the limestone bedding, which dips about 30° E. Only small patches of fluorspar are present in the north end of the pit, and none in the south end.

A shallow pit near the south end of the Lucky Boy No. 2 claim exposes an eastward-striking vertical vein of coarse light-green to almost colorless fluorspar. Cleavage fragments of fluorite as much as 2½ inches across can be obtained. The vein has a maximum width of 2 feet and is very irregular. The pit is 30 feet northeast of an outcrop of an intrusive body mapped as quartz monzonite by Butler. Small veins of fluorspar are exposed in other shallow pits, and have a maximum width of about 6 inches.

HUB, LADY BRYAN, MONTE CRISTO, MOSCOW, AND WILD BILL MINES AND MANASSA CLAIM

The Hub, Lady Bryan, Monte Cristo, Moscow, and Wild Bill mines (fig. 10) are old copper-lead-zinc-silver mines that were most active in the early 1870's. Butler (1913) has described the deposits of the Hub (p. 201–203), Lady Bryan (p. 203–204), Moscow (p. 201–202), and Wild Bill (p. 196–197) mines. Fluorite is associated with muscovite, magnetite, galena, sphalerite, pyrite, and chalcopyrite at the Hub mine; with muscovite, epidote, pyroxene, and tremolite at the Lady Bryan mine; and with garnet, diopside, magnetite, specularite, and sulfide ores at the Wild Bill mine.

The surface exposures at the Monte Cristo and Moscow mines and a tunnel at the Manassa claim were examined. At the Monte Cristo mine, fluorspar veinlets in limestone are exposed near the collar of the inclined shaft and about 100 feet north of it. The shaft follows an eastward-dipping fault between quartz monzonite on the east and limestone on the west.

The Moscow mine was being worked in June 1943 for sulfide ores containing galena, sphalerite, pyrite, and chalcopyrite. Medium-grained fluorite occurs as a gangue mineral with rhodochrosite and sericite. According to Stevens, the ore contains as much as 20 percent of CaF_2. Fluorite also occurs with a fibrous amphibole in altered dikes.

Medium-grained fluorspar occurs on the Manassa claim in a contact-metamorphic deposit in limestone and is associated closely with a fibrous amphibole and less closely with garnet. Little is known of the extent of the fluorspar zone, but it is probably very irregular. A fluorspar zone 2 to 3 feet wide was seen in the end of a 200-foot tunnel. The grade of the fluorspar in the tunnel is probably not more than 20 percent CaF_2. Little fluorspar is exposed on the surface.

QUARTZITE CLAIM

On the Quartzite claim (fig. 10) a vein of coarse-grained fluorspar is exposed in a shallow pit along a bedding plane in quartzite. The quartzite is probably a lens in the Harrington formation (Triassic). The bedding strikes N. 20° E. and dips 45° E. The fluorspar vein has a maximum thickness of about 2 feet, and about 10 tons has been mined from it. At the north end of the pit a little fluorspar is found in a fault that strikes east and dips 75° S. between the quartzite and a limestone north of it.

VIRGINIA AND NORTH VIRGINIA EXTENSION CLAIMS

Fluorspar occurs on the Virginia claims (fig. 10) in quartz-fluorite veins in an intrusive rock, probably a part of the southernmost body of quartz monzonite mapped by Butler.

In a 15-foot inclined shaft on the Virginia No. 2 claim, an irregular vein striking N. 30° W. and dipping 30° SW. is exposed. The maximum vein width seen is 2 feet, but Stevens states that part of the vein now mined out had a width of 4 feet. Probably about 50 tons was mined. One shipment from the shaft contained 73 percent of CaF_2.

On the Virginia No. 3 claim a shallow open cut about 40 feet long exposed an eastward-striking lens of fluorite, calcite, and quartz, which, according to Stevens, was as much as 6 feet wide and contained 60 to 70 percent CaF_2. The dip at the surface was about 20°, but steepened to about 40° at a depth of 5 feet; the lens narrowed at this depth. Probably about 50 tons of fluorspar has been mined. Only small fluorspar veinlets are exposed at the ends of the pit.

Fluorspar veinlets are found in the quartz monzonite on the Virginia No. 1 and North Virginia Extension claims, and, according to Stevens, also between the Virginia No. 1 and Virginia No. 2 workings and on the Virginia No. 4 claim.

DEPOSITS IN THOMAS RANGE DISTRICT, JUAB COUNTY

By M. H. Staatz, V. R. Wilmarth, and H. L. Bauer, Jr.

The Thomas Range fluorspar district in Juab County is one of the newest mining districts in the western United States. The first producing mine was opened by the Spor brothers in 1943, and large-scale mining began in 1948 after the discovery of the Blowout, Fluorine Queen, and Lost Sheep deposits; by September 1950 the district had produced 35,700 short tons of fluorspar. The abnormal radioactivity of the fluorspar has aided in the discovery of new deposits.

The Thomas Range is a northwest-trending group of mountains that includes Topaz Mountain, about 7,110 feet high. The fluorspar deposits are in the western branch of the range, called "Spor's Mountain" by Fitch, Quigley, and Barker (1949, p. 63-66). Spors Mountain is in Tps. 12 and 13 S., R. 12 W., Salt Lake principal meridian (fig. 1). It is about 6 miles long, and the highest peak is about 6,600 feet in altitude. The deposits are accessible by mine roads that connect with haulage roads on both sides of Spors Mountain (fig. 11). The haulage roads join the Callao-Delta road 57 miles west of Jericho, Utah, and 45 miles northwest of Delta, Utah.

In November 1943 A. E. Granger of the Geological Survey examined the Floride (Original Spor) mine, and in August 1944 the deposit was visited by W. R. Thurston and E. D. Washburn. The Spor mine has been revisited by members of the Geological Survey several times since then. No geologic maps of the Thomas Range have been published, but some mapping has been done by W. P. Fuller of the International Smelting & Refining Co., and by James Quigley of Centennial Development Co., Eureka, Utah. The district has been described by Fitch, Quigley, and Barker (1949). The fluorspar deposits were mapped and studied in August 1950 by M. H. Staatz and H. L. Bauer, Jr., and in September 1950 by V. R. Wilmarth and H. L. Bauer, Jr., as part of an investigation of uranium resources being carried out by the U. S. Geological Survey on behalf of the Atomic Energy Commission.

The writers are indebted to the mine operators for making available information and in giving other assistance. They are grateful to Mr. James Quigley for permission to use figure 11. Mr. Quigley and Mr. W. P. Fuller were generous in giving of their knowledge of the geology of the district in discussions of the Thomas Range deposits.

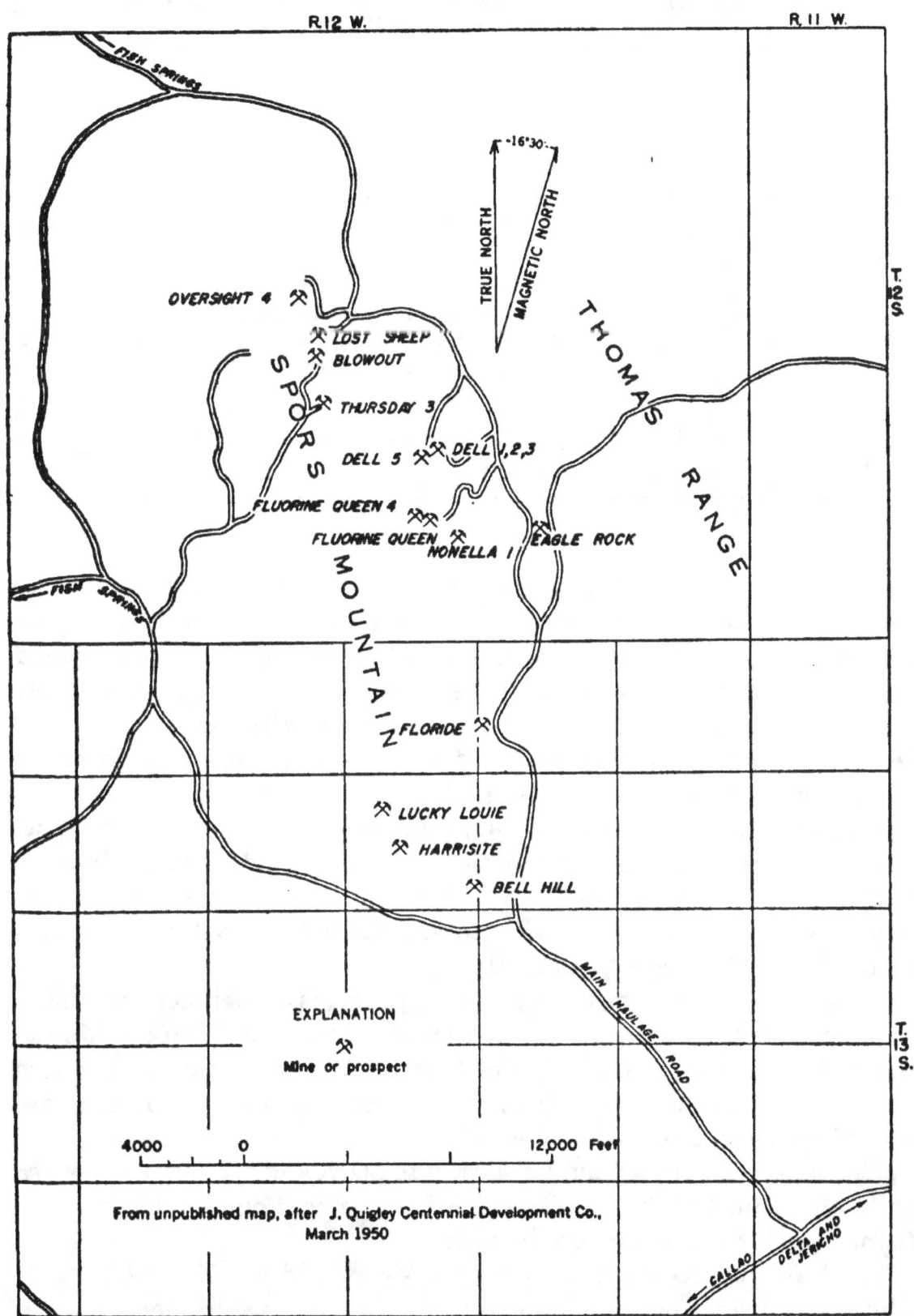

FIGURE 11.—Index map of the Thomas Range district, Juab County, Utah.

Fluorspar has been produced from 12 deposits on 8 properties through September 1950, as summarized in table 2. The material shipped generally contained 65 to 94 percent CaF_2, less than 2 percent silica, and less than 5 percent calcium carbonate. The chief impurity is clay; phosphorous and sulfur are generally absent.

TABLE 2.—*Fluorspar produced in the Thomas Range district, Juab County, Utah, through September 1950*

Property	Owner	Operators	Number of deposits that produced fluorspar	Production (short tons)
Floride	G. P. Spor, Chad Spor, Ray Spor.	G. P. Spor, Chad Spor, Ray Spor.	1	10, 285
Dell Nos. 1, 2, and 3.	Ward Leasing Co	Ward Leasing Co	3	7, 000
Dell No. 5	T. A. Claridge, Earl Willden, Albert Willden, Lafe Morley.	G. P. Spor, Chad Spor, Ray Spor.	1	275
Fluorine Queen	W. E. Black, F. B. Chesley	W. E. Black, F. B. Chesley	2	6, 770
Blowout	T. A. Claridge, Rex Claridge	T. A. Claridge	1	3, 000
Lost Sheep	Albert Willden, Earl Willden	Albert Willden, Earl Willden	1	6, 400
Bell Hill	H. J. Ruthlford, D. W. Searle, C. D. Searle, H. E. Searle.	Chad Spor, G. P. Spor, Ray Spor.	2	1, 870
Thursday No. 3	W. L. Hamblin, G. D. Hamblin, A. W. Hamblin.	W. L. Hamblin, G. D. Hamblin, A. W. Hamblin.	1	100
Total				35, 700

GENERAL GEOLOGY

The Thomas Range is composed of Paleozoic sedimentary rocks and Tertiary volcanic rocks. The eastern part of the range is made up principally of rhyolite tuffs and flows that are capped by a fine-grained gray, massive topaz-bearing rhyolite. The western part of the range, Spors Mountain, is underlain by an older quartzite, several hundred feet thick, and a thick sequence of younger dolomite. These rocks have been complexly step-faulted by steeply dipping normal and reverse faults, and intruded by plugs and dikes of rhyolite porphyry. Uraniferous fluorspar deposits occur in dolomite and intrusive breccia in the western part of the Thomas Range. Thin conglomerates, deposited along the shores of Pleistocene Lake Bonneville, are found locally along the base of Spors Mountain.

The quartzite forms prominent outcrops and is a thick-bedded rusty-weathering rock. It crops out along the eastern side of Spors Mountain and, according to W. P. Fuller (personal communication) extends eastward to a fault of several thousand feet displacement on the east side of Spors Mountain.

The dolomite ranges from light gray to black and from fine to coarse grained. Some beds contain as much as 20 percent chert. The writers found fossils of Silurian age in them.

The Paleozoic rocks strike N. 20° to 40° E. and dip 25° to 45° NW. Three sets of steep normal and reverse faults form a step-block pattern that causes many repetitions and omissions of individual beds. Fuller (personal communication) has noted north-striking and northeast-striking sets, and the writers also noted an east-striking set at the Bell Hill and Harrisite properties. The displacements range from a few feet to several thousand feet. Most of the faulting took place before the emplacement of the volcanic rocks, but postemplacement faults of small displacement are known.

Plugs and dikes of intrusive breccia commonly made up of rhyolite porphyry fragments in an aphanitic groundmass intrude the dolomite. The rhyolite porphyry is commonly white and contains phenocrysts of quartz and orthoclase one-eighth inch long in an aphanitic groundmass. The groundmass of unaltered rhyolite porphyry fragments is chiefly orthoclase and a few percent of quartz and euhedral topaz. Commonly the orthoclase of the rhyolite porphyry fragments associated with fluorspar deposits is altered to clay minerals and sericitized, and topaz is absent from the altered rock. The intrusive breccia does not crop out at the surface, but the shapes of the intruded bodies in most places are outlined by outcrops of the more resistant adjacent dolomite. In some places the soft altered intrusive breccia is veined by fluorite. The intrusive rock is presumed to be of Tertiary age because similar extrusive rocks east of the Thomas Range, 7 miles southwest of Levan, Utah, have been dated thus (Muessig, 1951, p. 234).

All the fluorspar deposits discovered so far are epithermal pipelike bodies or veins and occur in dolomite or rhyolite porphyry but not in quartzite. The deposits are either vertical or steeply dipping. Circular or oval pipelike bodies have been the source of about 95 percent of the fluorspar mined in the district. These bodies range in size from 15 feet in diameter to 176 feet long and 103 feet wide (east pipe, Fluorine Queen, fig. 14). The pipes commonly are associated with faults. The margins of the fluorspar bodies are sharp, and soft fluorspar is commonly in contact with unaltered dolomite (pl. 7A). At the Blowout property (fig. 12) the fluorspar pipe is partly in contact with an altered intrusive breccia plug; at the Harrisite property (pl. 8) a fault that cuts dolomite and altered basalt contains fluorspar.

Irregular pipelike bodies have ragged margins, and veinlets of fluorite extend outward into dolomite (pl. 7B). Blocks of dolomite as much as 6 feet in diameter commonly are included in the fluorspar bodies, as in pit No. 1 of the Bell Hill property. An east-striking fault along the west wall of pit No. 2 on the Bell Hill property (pl. 5) is marked by well-developed slickensides, but in most irregular deposits no fault surfaces are evident. These deposits probably formed close to faults along related zones of closely spaced fracturing in the dolomite.

The irregular pipelike bodies range in size from 10 feet wide and 25 feet long (trench 3, Bell Hill property, pl. 5) to 25 feet wide and 145 feet long (pit No. 2, Bell Hill property, pl. 5).

Veins of fluorite are common, but only at the Thursday No. 3 property (fig. 17) has a vein yielded fluorspar in commercial quantities. The veins range in size from a few inches wide and a few feet long to 14 feet wide and 246 feet long. Most of the veins are along fractures in dolomite, but a few are localized along faults in rhyolite porphyry dikes, as in trench No. 3, Harrisite property (pl. 8) and on the Thursday property (fig. 17).

White to deep-purple fluorite constitutes 65 to 95 percent of each deposit. It is soft and closely resembles a white to purple clay. Clay minerals, dolomite, calcite, carnotite, quartz, and opal are the other minerals visible. The fluorspar of most properties contains 2 percent or less silica, but some contains at least 20 percent. The calcite and dolomite contents are low except in a few deposits, such as at the Bell Hill and Harrisite properties. The Geneva Steel Co. reported (Fitch, Quigley, and Barker, 1949, p. 65–66) that the first ore shipped from the Floride claim contained 95 percent CaF_2 and 1 percent SiO_2, no phosphorous, and no sulfur. The first 35 cars shipped from this property averaged more than 90 percent CaF_2 and less than 2 percent SiO_2. About 100 railroad cars of ore were shipped from the Lost Sheep property; it assayed more than 90 percent CaF_2 and less than 1 percent SiO_2. One carload lot was reported to have the following analysis: 94.90 percent CaF_2, 0.44 percent SiO_2, 1.12 percent CaO, 0.32 percent MgO, 0.012 percent S, and 4.5 percent H_2O. The first 35 carloads of fluorspar from the West pipe on the Fluorine Queen property contained 70 to 90 percent CaF_2 and less than 2 percent SiO_2; the lower grade ore came from near the surface.

All the fluorspar deposits are abnormally radioactive, but generally no uranium minerals can be seen. Assays of 45 fluorspar samples ranged from 0.005 percent to 0.33 percent uranium. Samples from three deposits on the Bell Hill, Harrisite, and Eagle Rock properties contained more than 0.1 percent uranium. Carnotite was noted only in the fluorspar at the Eagle Rock property, the Floride property, and in the West pipe at the Fluorine Queen property. At the Bell Hill and Harrisite properties carnotite is found in the adjacent country rock but not in the fluorspar.

SUGGESTIONS FOR PROSPECTING

The most favorable place to find fluorspar deposits is in areas adjacent to faults and intrusive breccia pipes. A detailed geologic map of the faults and intrusive bodies in the area would show most of the places favorable for finding fluorspar deposits.

Most of the fluorspar is radioactive, and traversing the surface with Geiger-Mueller counters should assist in delineating favorable areas for prospecting. Oval or circular areas surrounded by dolomite outcrops are favorable sites for finding concealed, intrusive breccia or fluorspar bodies; many of these areas are shallow depressions filled with dolomite and chert float and can be prospected by bulldozing.

Float is an invaluable aid in prospecting in this district. The fluorspar float is very soft and is rarely found far from the deposits.

Commonly it is included in the debris surrounding anthills, and rabbit and other animal burrows above the deposits.

In the southern part of Spors Mountain, where the relief is low and overburden is thick, the best method of prospecting in absence of float, is to make a geologic map of the area and to cut with bulldozer trenches the favorable areas that are revealed by the mapping.

DESCRIPTION OF INDIVIDUAL DEPOSITS

BELL HILL PROPERTY

The Bell Hill property (fig. 11) is on a hill about 200 feet high at the southern end of Spors Mountain. It is in sec. 10, T. 13 S., R. 12 W., Salt Lake principal meridian. The Bell Hill group of claims was staked July 17, 1949 by H. J. Ruthiford, D. W. Searle, C. D. Searle, and H. E. Searle. The claims have been leased by G. P. Spor of Delta, Utah, who was actively mining them in September 1950. Thirty-four cars of fluorspar, averaging 55 tons each, were mined and shipped between July and September of 1950. This material had an average CaF_2 content of more than 70 percent.

Workings on the Bell Hill property (pl. 5) consist of three stripped areas and two open pits. Pit No. 1, from 10 to 26 feet deep, is on the largest fluorspar body. About one carload of fluorspar a day was being shipped from this pit in September 1950.

The fluorspar bodies occur in gray cherty dolomite under a cover of dolomite float. Outcrops of the dolomite are not conspicuous, and poor exposures make it difficult to recognize faults in the homogeneous dolomite. A fault is well exposed along the north side of pit No. 2: it strikes N. 76° E. and dips 65° SE. Grooves and slickensides on the fault plane plunge 54° S. 65° E. The other fluorspar bodies on the property probably formed along faults also, but evidence of the faults has been destroyed by the introduction of the fluorite.

At least five fluorspar deposits are exposed in the workings at the Bell Hill property. Trench No. 1 contains two 6-foot veins and a 6-inch stringer of fluorspar, trench No. 3 exposes a 3- and a 9-foot vein of fluorspar. The fluorspar presumably has replaced the dolomite along faults or small shears. The deposits have a general linear trend but in detail are extremely irregular. Four of the bodies trend nearly east, and the fifth strikes northeast. All, except the fluorspar body in pit No. 2, are essentially vertical (pl. 5). In pit No. 1 a boulder of dolomite, 6 feet in diameter, was completely surrounded by fluorspar. The shape of the deposit in the northwest-trending segment of the deposit in pit No. 2 is extremely irregular (pl. 7B). The fluorspar extends outward into the dolomite along a series of parallel shears, and the edge of the body has a serrate appearance.

Two small bodies, or a branching deposit, are exposed in trench No. 3. The east exposure is 24 feet long and as much as 10 feet wide; the west exposure is 9 feet long and as much as 2.5 feet wide.

The fluorspar in pit No. 2 averages 25 feet in width and 145 feet in length, and is very irregular, especially in the northwest-trending part. It was mined to a maximum depth of 15 feet.

The largest deposit is in pit No. 1 and is split by a mass of dolomite. The body has an exposed length of 130 feet and a maximum width of 45 feet at the junction of the split. Mining was in progress at a maximum depth of 26 feet in this pit in September 1950. Carnotite was found as a coating on the gray dolomite adjacent to the fluorspar body in pit No. 1.

BLOWOUT PROPERTY

The Blowout property (fig. 11) is in the south-central part of T. 12 S., R. 12 W., Salt Lake principal meridian. The claim was located on May 10, 1948 by T. A. Claridge and Rex Claridge of Delta, Utah. The mine is accessible over a dry-weather road along the western side of Spors Mountain that connects with the main haulage road 3 miles to the south. A total of 3,000 short tons of fluorspar, averaging 75 percent CaF_2, has been produced from a pipe on the property (fig. 12).

The Blowout deposit has been developed by a large opencut and a crosscut 240 feet below the opencut. The opencut trends N. 70° W. and is 150 feet long. It averages 30 feet wide and is 57 feet deep at the western edge of the cut. At 500 feet from the portal the crosscut intersects the deposit exposed in the opencut. The crosscut was driven by the owners of the Lost Sheep and the Blowout mines for use jointly as a haulage adit.

At the Blowout property intrusive breccia intrudes dolomite (fig. 12). The dolomite beds strike N. 43° E. and dip 34° to 43° NW. The opencut contains exposures of the contacts between dolomite, intrusive breccia, and fluorspar. The intrusive breccia is now a hematite-stained clay containing calcite. Fragments of this rock found on the dump show rounded phenocrysts of smoky quartz and remnants of feldspar crystals in an altered, hematite-stained matrix. The dimensions of the igneous body could not be determined because it is partly covered by dump material. In spite of the apparent relationship of the intrusive breccia cutting across the fluorspar body, as shown in figure 12, field study indicates that the intrusive breccia is earlier.

The fluorspar is exposed in the pit as an irregular pipelike deposit that averages about 30 feet wide and 110 feet long. At the northern edge of the opencut a fissure-type extension of the deposit, 17 feet wide, extends for 37 feet along the northwestern edge of the igneous body. Fluorspar is exposed to a depth of 57 feet in the opencut and the cross-

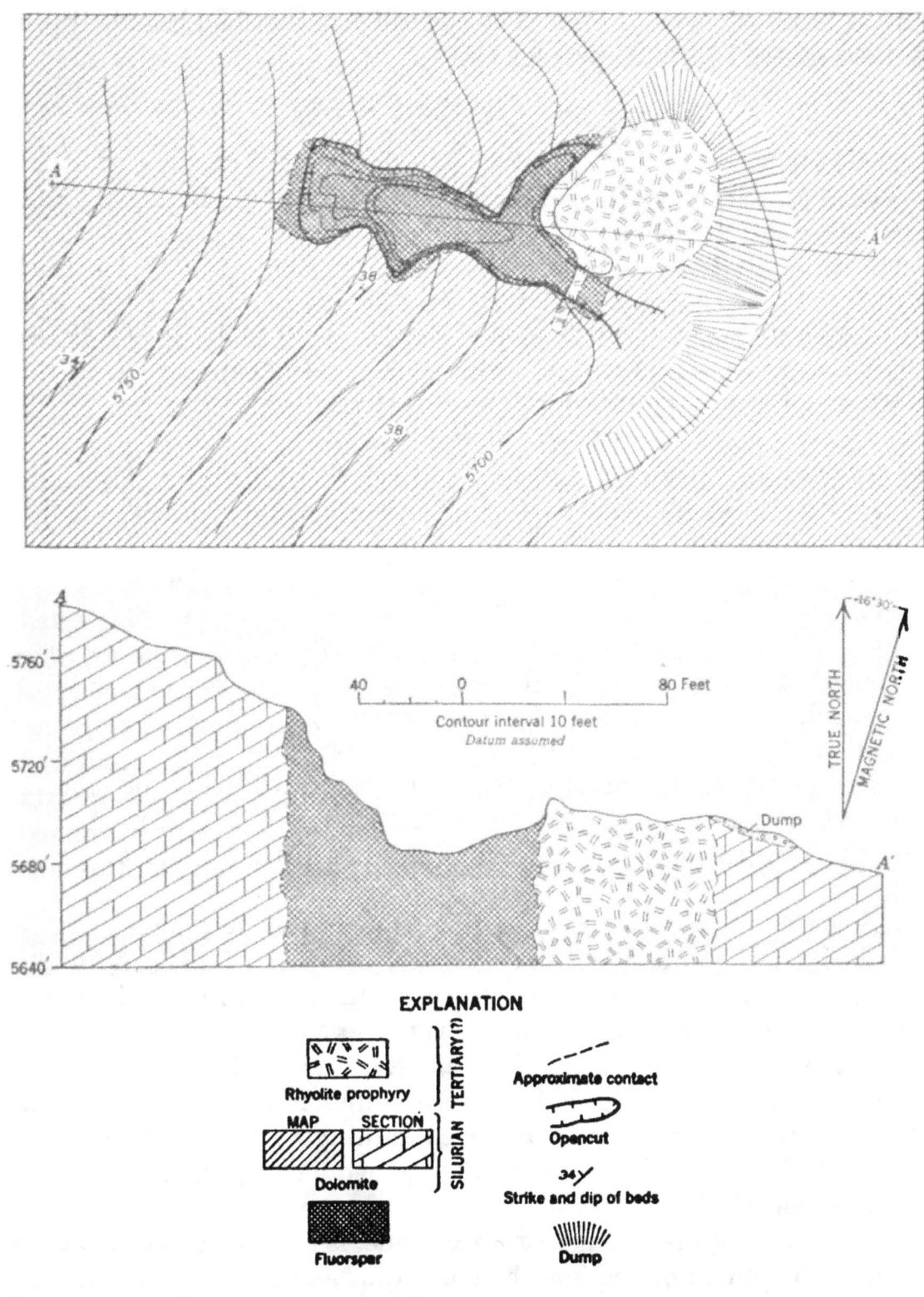

EXPLANATION

Rhyolite prophyry

MAP   SECTION

Dolomite

Fluorspar

Approximate contact

Opencut

Strike and dip of beds

Dump

Geology and topography by V. R. Wilmarth and H. L. Bauer, Jr.,
September 1950

FIGURE 12.—Map and section of the Blowout property, Juab County, Utah.

cut intersects the deposit at 240 feet below the level of the opencut; the size of the deposit at that depth is not known. A 6-inch zone of dolomite at the contact has been altered to fine-grained sugary dolomite containing thin stringers of calcite. The fluorspar is white, pale violet, or dark purple. It ranges from a soft white to a hard, dark-purple, boxwork type of ore.   The carload lots that had been shipped by 1950 assayed 75 to 80 percent CaF_2.

DELL NOS. 1, 2, AND 3 CLAIMS

The Dell Nos. 1, 2, and 3 claims (see fig. 11) are on the east slope of Spors Mountain, in the southeast part of T. 12 S., R. 12 W., Salt Lake principal meridian.

The original location was made in the spring of 1947 by Earl Willden, T. A. Claridge, and Lafe Morley, all of Delta, Utah, who drove an adit about 100 feet long toward a small body of ore at the northern end of the property.   The claims were sold in 1948 to the Ward Leasing Company, who continued this adit about 15 feet to the deposit; this company also drove a haulage adit, 933 feet long, S. 60° W. from a point 1,500 feet southwest of the small deposit.   This adit prospected a large fluorspar body, known as the main or western body.

The country rocks at the Dell Nos. 1, 2, and 3 deposits are dolomite, quartzite, and plugs of intrusive breccia.   The dolomite is a dark-gray, crystalline rock containing much chert and is commonly brecciated along fault zones.   A rhyolite porphyry intrusive body crops out near the center of a large intrusive breccia plug.   This rock has phenocrysts of orthoclase and smoky quartz in a siliceous groundmass partly replaced by fluorite.   All gradations can be observed from unaltered rhyolite porphyry to rhyolite porphyry in which only the quartz phenocrysts remain unaltered in a completely altered groundmass. Most of the orthoclase is highly kaolinized; a little sericite also was formed.

The main, or western body, 30 feet wide and 50 feet long, is oval in cross section and is intersected 200 feet below the surface by a 50-foot raise from the 933-foot long adit.   This deposit is in dolomite and ends against the underlying quartzite exposed at the top of the raise.   A smaller body, 32 feet in diameter, was cut 40 feet from the portal of the long haulage adit.   Both bodies have been mined out. About 7,000 short tons of fluorspar containing 80 percent CaF_2 was produced.   The material as shipped contained 3 to 4 percent SiO_2 and 1 to 2 percent $CaCO_3$.

A third body, 10 feet long and 8 feet wide, is exposed about 800 feet north of the portal of the long haulage adit, and is intersected about 35 feet below the surface.   The deposit was stoped for 25 feet above and 25 feet below the adit level.   The deposit pinched out against quartzite at the bottom of this stope.   About 100 tons of fluorspar was produced.

An unmined, siliceous fluorspar vein occurs in the intrusive breccia plug near the portal of the 115-foot adit.   This fluorspar vein is 40 feet long and averages 1.5 feet in thickness.

DELL NO. 5 PROPERTY

The Dell No. 5 property (fig. 11) is in a small gap on the crest of Spors Mountain, in the southeast part of T. 12 S., R. 12 W., Salt Lake principal meridian. The workings are reached from the main haulage road by a private road about 1 mile long. The Dell No. 5 claim was located originally by Earl Willden, Albert Willden, T. A. Claridge, and Lafe Morley. It was leased for a period by the Spor brothers, who made the road, did the development work, and took out about 275 tons of fluorspar.

Two fluorspar deposits are on the property. The one nearest the road (the south body) was mined from a 60-foot adit that intersected the deposit about 25 feet below the surface. This body was stoped for 15 feet above and below the adit level. The other deposit (the north body) is partly exposed by bulldozer workings and is prospected by an adit about 15 feet long.

The country rock is a dark-gray dolomite containing some chert. A rhyolite porphyry plug, about 25 feet in diameter, cuts the dolomite. Phenocrysts of white, euhedral orthoclase, as much as one-sixteenth of an inch long, and quartz, as much as one-eighth of an inch long, make up about 35 percent of the rhyolite porphyry; the rest is an aphanitic white groundmass consisting chiefly of slightly kaolinized orthoclase and some clear topaz. The topaz forms euhedral crystals enclosed in both quartz and orthoclase and appears to have been the earliest mineral to crystallize.

The two fluorspar bodies are about 230 feet apart. The south body is about 15 feet in diameter and contains friable white to purple fluorite. The north body is incompletely exposed but has a minimum size of 15 feet by 20 feet in plan. The fluorspar is in part powdery but contains many hard ribs of siliceous material. It has been reported that a chip sample from the walls of the small adit contained 91 percent CaF_2 and 0.97 percent silica.

EAGLE ROCK PROPERTY

The Eagle Rock property (fig. 11) is on the east side of a low range of hills about 300 feet high, which are in the valley separating the western and eastern parts of the Thomas Range, in the southeast part of T. 12 S., R. 12 W., Salt Lake principal meridian. The Eagle Rock property consists of the Eagle Rock and Eagle Rock No. 1 claims. These claims were located on August 15, 1948, by T. A. Claridge, Rex Claridge, Albert Willden, and Earl Willden, all of Delta, Utah.

On the Eagle Rock property (fig. 13) a fluorspar vein is exposed in a trench and a pit about 250 feet above the valley bottom and the nearest road. The trench is 30 feet long, 4 feet wide, and a maximum of 12 feet deep. The pit is 4 feet square, and is 20 feet southwest of the trench. Five other prospect pits have been made in intensely altered dolomite.

The country rock on the Eagle Rock claims consists of a fine-grained, dense, gray dolomite containing a fine network of chert. The dolomite strikes N. 43° E. and dips 30° NW. In the central part of the property a depression is covered by slope wash (fig. 13); similar smaller depressions are found elsewhere on the property. The general circular shape of the depressions, the brecciated and silicified dolomite in the surrounding outcrops, and the presence of calcite-rich, cream to reddish claylike material in the pits suggest that the depressions formed in hydrothermally altered dolomite above intrusive plugs.

The fluorspar body exposed on the Eagle Rock property (fig. 13) is extremely irregular in size and shape. It is 4 feet wide at a depth of 6 to 12 feet in the middle of the trench and is absent at a depth of 2 feet. At 14 feet from the north end of the trench, half a foot of fluorspar is exposed, and at 24 feet the vein is 4 feet wide; several small patches of fluorite were found between these two exposures. The deposit is vertical and in the trench and pit has a minimum length of 40 feet.

The fluorspar is dark purple and is mixed with at least 30 percent of silica that forms a fine boxwork structure. This material is unusually hard and resistant in contrast to the powdery fluorspar common in the district. The fluorspar is enclosed in white soft limy material that is probably calcite resulting from alteration of the dolomite. Carnotite coats the fluorite in some places. This body is one of the few in the district that contain visible uranium minerals with the fluorite.

FLORIDE (ORIGINAL SPOR) MINE

The Floride (Original Spor) mine (fig. 11) was the first property operated in the Thomas Range district. It is on the eastern slope of Spors Mountain, at an altitude of about 5,300 feet, in the SW¼ sec. 2, T. 13 S., R. 12 W., Salt Lake principal meridian. The mine is easily accessible over a steep road that leads westward from the main haulage road. The Floride property was staked on January 1, 1941, by Chad Spor, G. P. Spor, and Ray Spor, who operated the mine intermittently between 1941 and 1950. The total production of fluorspar through 1950 is 10,285 short tons that averaged 80 to 85 percent CaF_2.

The Floride deposit has been mined by shrinkage methods from two haulage adits (pl. 6). The upper adit is about 130 feet long and intersects the deposit 40 feet below the surface. The lower adit is about 275 feet long and cuts the deposit about 80 feet below the surface. On this level, two crosscuts have been driven. The crosscut to the south is 70 feet long and was driven in fluorspar for a distance of 30 feet.

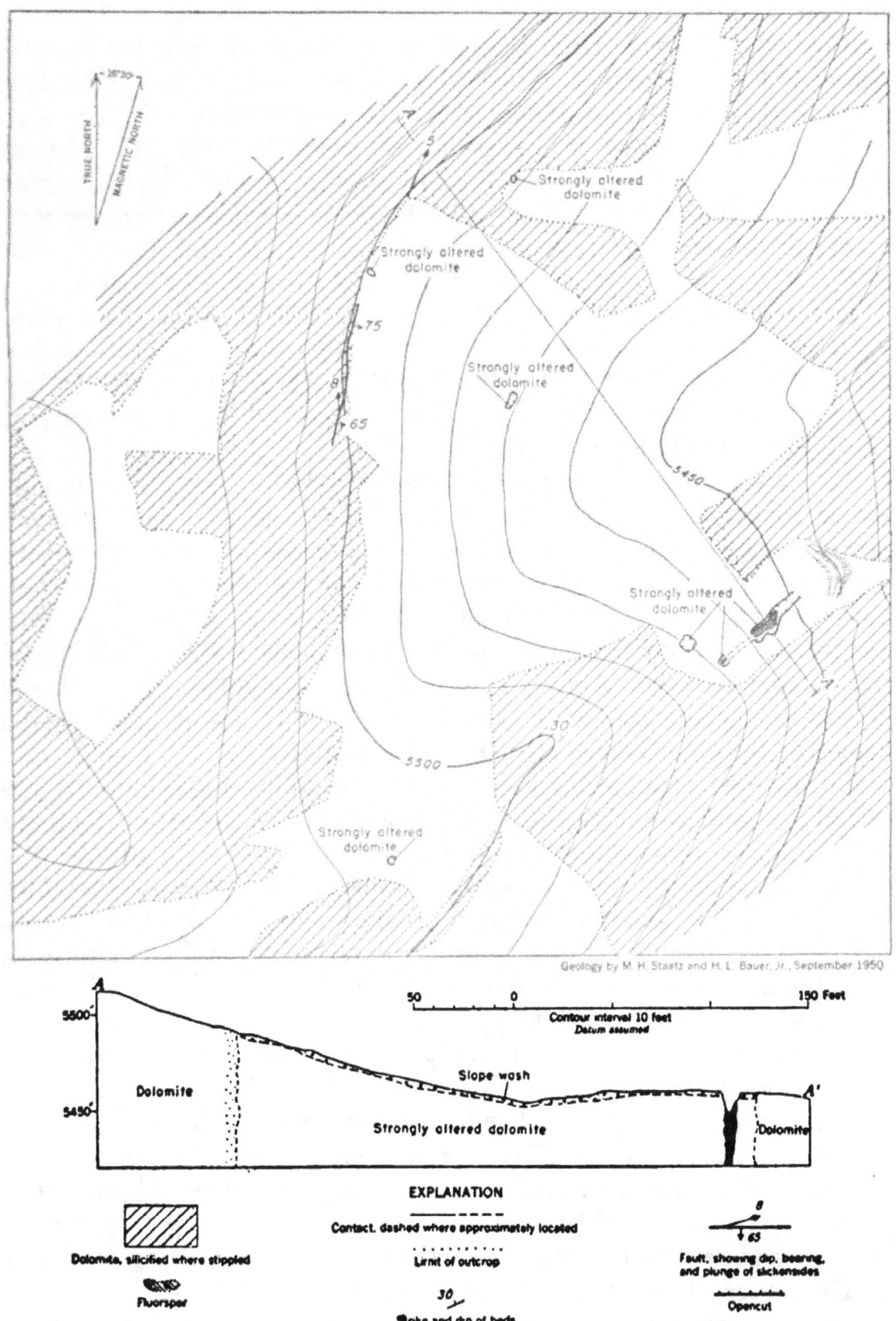

FIGURE 13.—Map and section of the Eagle Rock property, Juab County, Utah.

The other crosscut extends west to the edge of the body and is 30 feet long. A shaft inclined at 60° has been sunk from the surface to the lower adit level. The glory hole is an elongate, northwest-trending cut that is 60 feet long and 40 feet wide at the surface. An inaccessible vertical shaft extends from the bottom of the glory hole for an estimated 40 feet in fluorspar. Several small prospect pits have been dug south of the opencut on extensions of the fluorspar body.

The Floride deposit is a pipelike replacement body (pl. 7A) in dolomite near a northeast-trending fault that places the older quartzite against the younger dolomite. The dolomite is a dense, dark-gray to black, thick-bedded rock that strikes N. 32°–38° E. and dips 26°–32° NW. The quartzite strikes N. 30° E. and dips about 29° NW.

The major structure in the vicinity of the Floride mine is a normal fault that strikes N. 85° E., dips 55°–60° SE., and has a vertical displacement of about 200 feet. The quartzite was not fractured during the faulting, whereas a breccia zone as much as 10 feet wide occurs in the dolomite at the contact with the quartzite.

Two sets of fractures were observed in the lower adit. The strongest set strikes N. 30°–35° W. and dips 70°–90° NE. Commonly the second set, consisting of northeast-trending fractures, is filled by white coarsely crystalline calcite veins, as much as 1 foot thick. The fractures generally are not mineralized with fluorite and commonly contain caverns as wide as 1 foot. In the upper adit level a small fault forms the contact between altered and unaltered dolomite. The altered dolomite contains abundant calcite and limonite stains, and can be traced to the glory hole where the zone disappears.

The fluorspar body is irregular in outline and has a maximum length of 100 feet and an average width of 40 feet. The footwall of the body dips southeast approximately parallel to the fault; the hanging wall is essentially vertical. The irregular shape of the deposit, the relict bedding in the fluorspar adjacent to the dolomite, and the veinlike stringers of fluorspar along bedding planes in the dolomite indicate that the fluorspar is a replacement body in the dolomite. Calcite occurs as thin stringers and as pockets in the dolomite adjacent to the fluorspar and makes up 50 percent of the rock in some places.

The fluorspar is extremely variable in color and composition, ranging from a dark-purple, highly siliceous material to soft white powdery material that is 95 percent CaF_2. For the most part, it is fine-grained, white to pale purple and in carload lots assays 85 percent CaF_2. Some of the dark-purple highly siliceous material contains finely crystalline carnotite as radiated aggregates and as thin veinlets filling fractures.

Exposures of fluorspar-cemented breccia are numerous in trenches near the workings and in gullies to the south. Here, the dolomite shows no displacement, although some of the fluorspar-bearing zones are as much as 30 feet long. Evidently zones of shearing rather than simple faults were mineralized.

FLUORINE (FLUERIN) QUEEN PROPERTY

The Fluorine (Fluerin) Queen property is about 1,800 feet above the valley bottom to the east, in the southeast part of T. 12 S., R. 12 W., Salt Lake principal meridian. Two large fluorspar pipes, the west pipe and east pipe, about 100 feet below and on either side of the crest of Spors Mountain, are reached by a steep mountain road along the east face of the mountain (fig. 11). The Fluorine (Fluerin) Queen claim was located in March 1948 by W. E. Black and F. B. Chesley of Delta, Utah. About 6,770 tons of fluorspar were mined at this property from 1948 to 1950.

The west pipe (fig. 14) has been worked from an irregular open pit, 105 feet long and about 25 feet wide. The pit was from 28 to 45 feet deep in September 1950 but was not being worked. The east pipe has been stripped of overburden, and a cut 80 feet by 60 feet has been made. The pit is 15 feet deep on the west side.

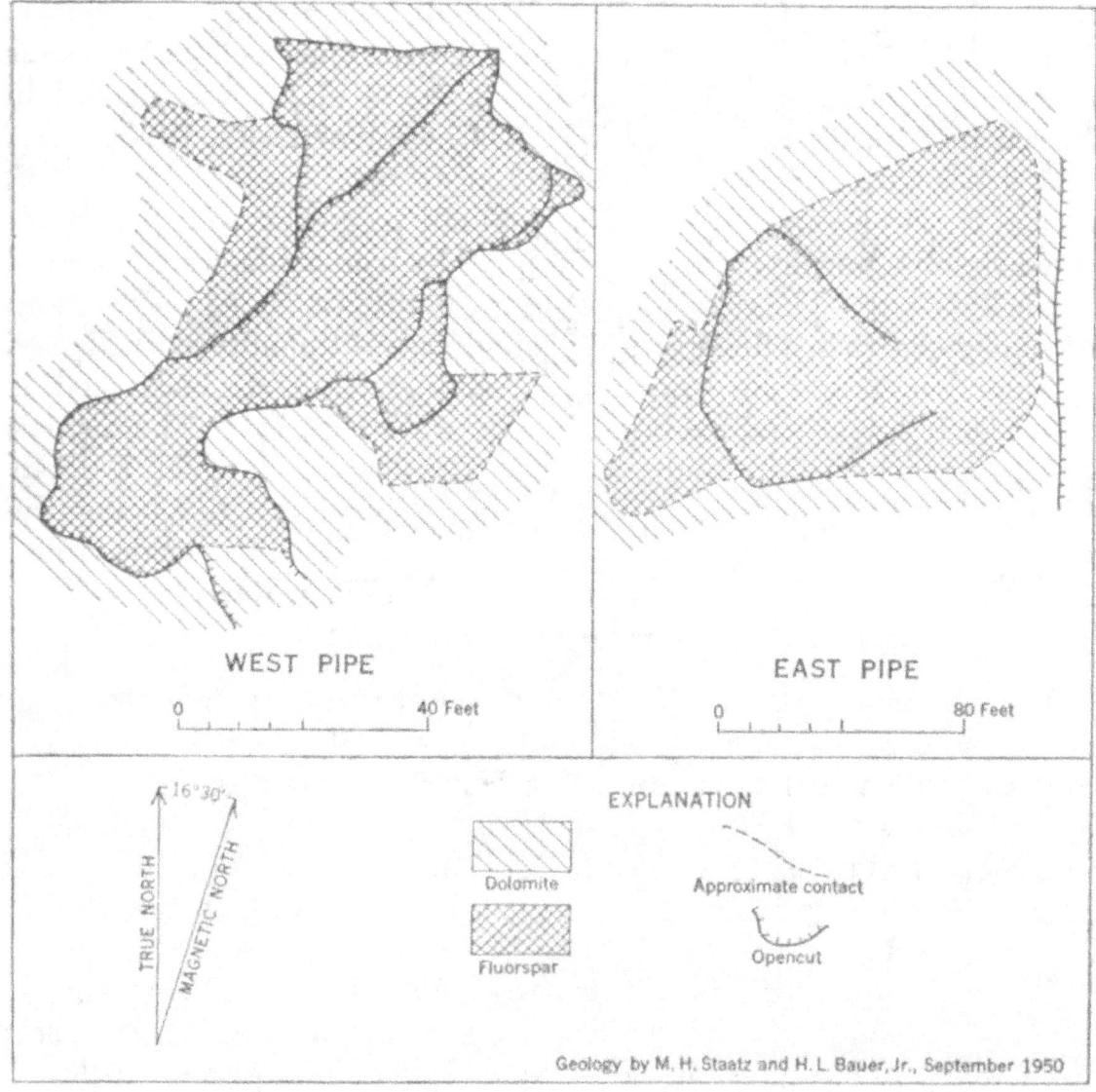

WEST PIPE

0   40 Feet

EAST PIPE

0   80 Feet

16°30'

TRUE NORTH

MAGNETIC NORTH

EXPLANATION

Dolomite

Fluorspar

Approximate contact

Opencut

Geology by M. H. Staatz and H. L. Bauer, Jr., September 1950

FIGURE 14.—Outlines of fluorspar pipes, Fluorine Queen property, Juab County, Utah.

Both fluorspar deposits are in dark-gray dolomite. The west pipe is irregular in shape, and walls dip steeply. The contact of the fluorspar and dolomite is sharp; the dolomite is not altered. This fluorspar body has a maximum exposed length of 105 feet and a maximum width of 80 feet. The east pipe is more regular in shape and is oval in plan. It has a maximum exposed length of 175 feet and a maximum width of 103 feet. Not enough development work has been done to determine the downward extension of this body.

The fluorite in both deposits is white to dark purple, soft, and friable. Carnotite is visible in a streak of hard fluorite, about 40 feet long and 1 foot wide, in the middle of the west pipe.

FLUORINE QUEEN NO. 4 CLAIM

The Fluorine Queen No. 4 claim (fig. 11) is on the southern slope of a southwestward-trending ridge that rises about 1,200 feet above the valley floor. The claim is on the west slope of Spors Mountain in the southeast part of T. 12 S., R. 12 W., Salt Lake principal meridian. It is reached from the west pit on the Fluorine Queen claim by means of a rough trail. The Fluorine Queen No. 4 claim was located in March 1948 by W. E. Black and F. B. Chesley of Delta, Utah.

A dense dark-gray cherty dolomite crops out in a prominent cliff, strikes N. 36° E., and dips 35° NW. Small fragments of a highly altered rhyolite porphyry, stained deep red by iron oxide, were observed in the fluorspar. The rhyolite porphyry contains rounded phenocrysts of smoky quartz in a fine-grained groundmass of limonite-stained clay.

Fluorspar occurs in a northwest-trending, elongate body that is about 90 feet long and averages 20 feet wide (fig. 15). The deposit has been prospected by four small pits, the deepest of which is about 8 feet. The body underlies a depression as much as 6 feet below the outcrops of the adjacent dolomite. The fluorite is white to pale violet. The matrix of the fluorite contains abundant limy material.

HARRISITE PROPERTY

The Harrisite property (fig. 11) is near the base of some low hills at the southern end of Spors Mountain in sec. 10, T. 13 S., R. 12 W., Salt Lake principal meridian. The Harrisite group of claims was located on May 10, 1949, by E. D. Harris, E. T. Harris, Rex Harris, and Mark Harris of Delta, Utah. No fluorspar had been produced from this property by September 1950, but according to the Harris brothers the deposit contains 60 to 65 percent CaF_2.

The main workings (pl. 8) are two bulldozer trenches (trenches No. 1 and 2) and a narrow hand-dug trench (trench No. 3). Trench No. 1, about 70 feet long and 15 feet wide, is in dolomite on the hillside. Trench No. 2, which measures 220 by 30 feet, partly exposes a fluor-

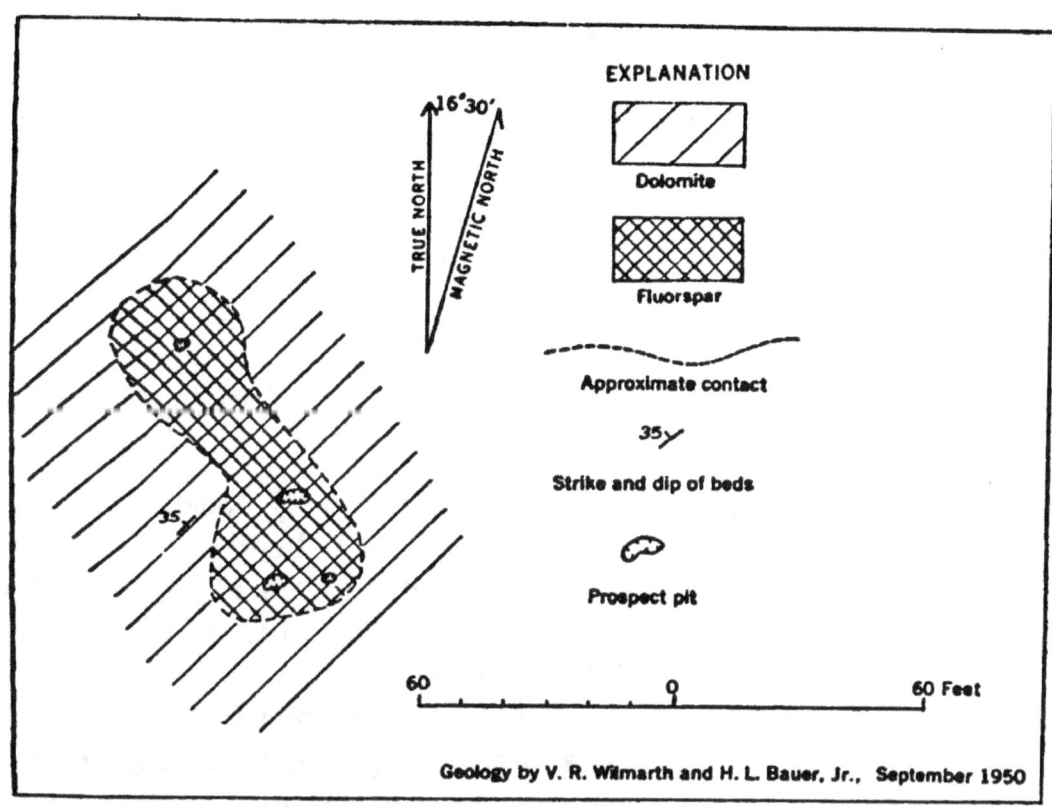

FIGURE 15.—Outline of fluorspar pipe, Fluorine Queen No. 4 property, Juab County, Utah.

spar body in the wash. Several small pits have been sunk in this trench. Trench No. 3, 40 feet long and 5 feet wide, has a maximum depth of 9 feet and exposes a small fluorspar vein at the fault contact of dolomite and altered rhyolite porphyry.

The rocks on the Harrisite property consist of several much-faulted dolomite units, conglomerate of the Lake Bonneville beds, basalt, and a small body of altered rhyolite porphyry. The dolomite strikes N. 6° W.–N. 29° E. and dips 25°–42° W.

.The southwestern part of the property is underlain by basalt. This rock is dark gray on fresh surfaces and deep brown on weathered surfaces. It consists of about 10 percent of dark-green, euhedral phenocrysts of pyroxene, as much as an eighth of an inch long, in a fine-grained aphanitic groundmass of plagioclase crystals, a little magnetite, and some glassy material. The basalt may be either a flow or an intrusive body and is presumed to be related to basaltic rocks south and east of Spors Mountain.

Basalt, which is now altered almost entirely to clay, crops out on the southern side of trench No. 3. The altered material retains its original texture. The altered basalt is cut by a small fault and many small parallel shears.

Six northeast-striking faults and one east-striking fault were mapped (pl. 8). Some small fluorite stringers exposed in the main wash may have formed on another fault parallel to the east-trending fault. The easternmost fault (pl. 8) offsets two recognizable units of the dolomite 64 feet horizontally. A second northeast-trending fault, dipping 74° SE., is exposed in trench No. 3 adjacent to altered basalt; it probably continues to the northeast. A third northeast-trending fault is inferred to be along the bottom of the wash, because of the omission of a considerable part of one dolomite bed in the southern part of the area. An outcrop of brecciated and silicified dolomite in the northeastern edge of the valley suggests faulting. The fourth fault is a strike fault that brings up units exposed in the eastern part of the map area. The displacement is at least several hundred feet. The fifth and sixth faults branch from this fault and appear to be later.

The fluorspar deposits at the Harrisite property consist of a thin veinlike body in trench No. 3 and large irregular bodies in trench No. 2. Powdery, white to dark-purple fluorite occurs along shears or small faults in dolomite and altered basalt. The thin, irregular body in trench No. 3 dips about 74° E., parallel to the dip of the adjacent fault. The attitude of the larger deposits in trench No. 2 is not definitely known and on plate 8 is shown as dipping vertically parallel to the adjacent fault. The deposits consist mainly of fluorite but contain some clay. Dolomite fragments are found in some places, especially in the fluorspar of trench No. 2. No uranium minerals are visible in the fluorite, but in trench No. 3 carnotite was noted along shear planes in the adjoining altered basalt.

The vein in trench No. 3 is 6 inches to 3 feet wide and is exposed for 20 feet along the trench. The irregular bodies in trench No. 2 are very poorly exposed. They cover an area at least 75 feet long, with a maximum width of 25 feet. Several stringers less than 1.5 feet wide extend about 100 feet to the west from the main body.

LOST SHEEP PROPERTY

The Lost Sheep property (fig. 11) is on the east side of Spors Mountain in the central part of T. 12 S., R. 12 W., Salt Lake principal meridian. The mine is on the side of a hill several hundred feet above the valley bottom to the east, and is reached by about one-quarter mile of private road from the north end of the haulage road on the east side of Spors Mountain. The Lost Sheep group of claims was staked May 10, 1948, by Albert Willden and Earl Willden of Delta, Utah.

Two fluorspar pipes, the main pipe and the west pipe, are on the Lost Sheep property. The main pipe is mined from an open pit 140 feet long and 60 feet wide (fig. 16). It is 54 feet deep on the west side. About 6,400 tons of fluorspar had been produced from this pipe through August 1950. The west pipe is about 800 feet southwest of the main pipe and crops out several hundred feet above it. An adit,

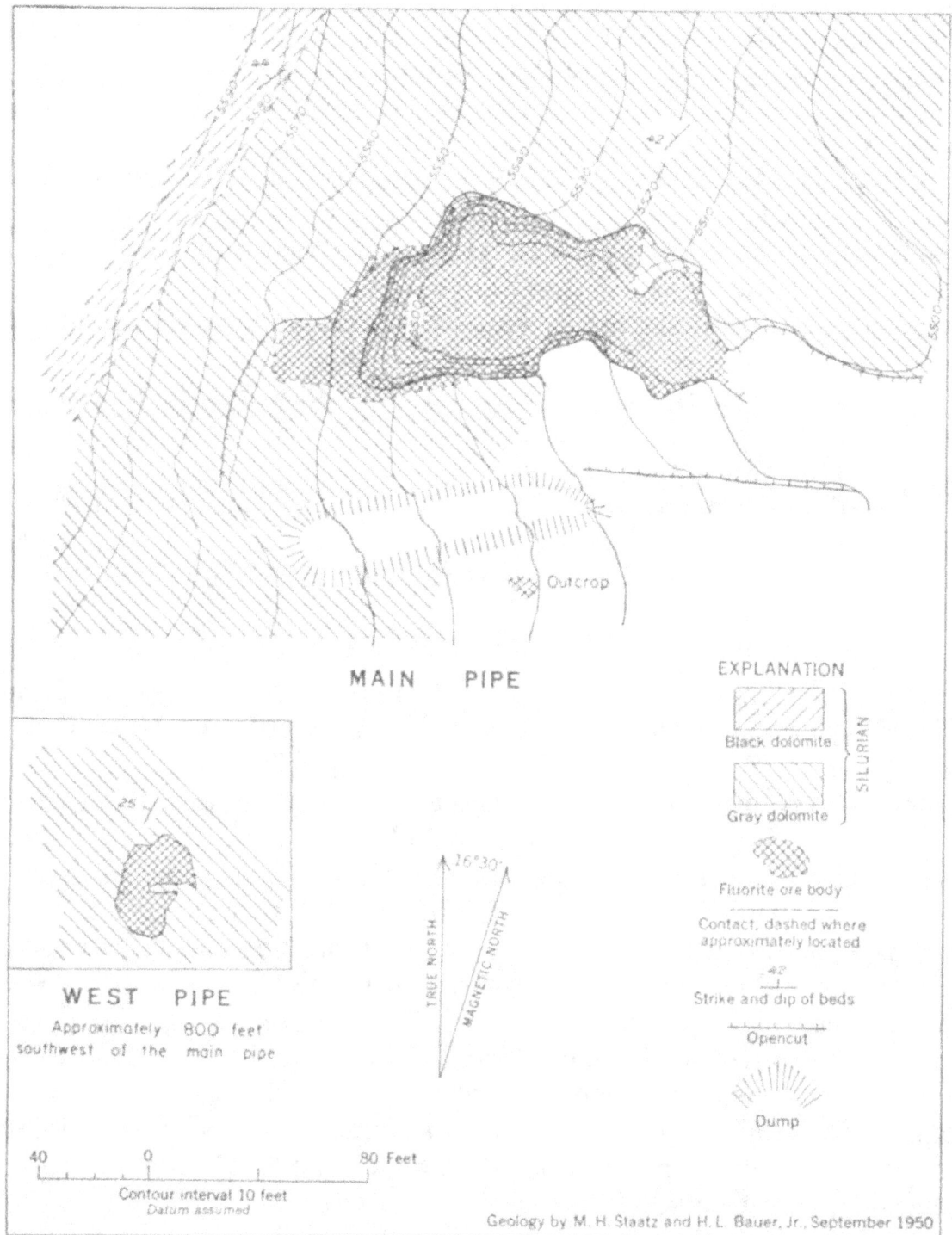

MAIN PIPE

EXPLANATION

Black dolomite
Gray dolomite
SILURIAN

Fluorite ore body

Contact, dashed where
approximately located

Strike and dip of beds

Opencut

Dump

WEST PIPE
Approximately 800 feet
southwest of the main pipe

Outcrop

TRUE NORTH   MAGNETIC NORTH   16°30'

40   0   80 Feet
Contour interval 10 feet
Datum assumed

Geology by M. H. Staatz and H. L. Bauer, Jr., September 1950

FIGURE 16.—Outlines of fluorspar pipes, Lost Sheep property, Juab County, Utah.

about 200 yards southwest of the main pipe, was driven jointly by T. A. Claridge and the Willden brothers. It is about 500 feet long and intersects the Blowout deposit 240 feet below the surface. A branch crosscut, started about 200 feet from the portal, is expected to reach the west pipe on the Lost Sheep property at a point about 200 feet below the surface.

The two fluorspar pipes on the Lost Sheep property are roughly oval in plan (fig. 16). The main pipe is 160 feet long, about 65 feet wide, and at least 54 feet deep; the walls dip steeply and may converge at depth.   The west pipe is 37 feet long and about 10 feet wide; the depth of the deposit is not known.   The main pipe is entirely in compact, light-gray dolomite that contains about 2 percent of chert. The dolomite is brecciated near the deposit in some places.   The beds strike N. 45°–50° E. and dip 42°–44° NW.

A dike or a large plug of intrusive breccia several hundred yards long is exposed in small pits and road cuts and in the adit about 400 feet southwest of the main pipe.   The adit starts in intrusive breccia, consisting of fragments of rhyolite porphyry in a matrix of similar material, and continues for 81 feet in this rock.   The fragments of rhyolite porphyry commonly are altered, but fresh surfaces show quartz phenocrysts and a few unaltered orthoclase crystals in a white aphanitic groundmass.   A 1½-inch vein of fluorite is exposed along the top of the adit near the portal.

Both pipes consist of soft, white to deep-purple, pulverulent fluorspar.   Carload lots of selected fluorspar are reported to contain as much as 95 percent fluorite.   No uranium minerals are visible.

LUCKY LOUIE PROPERTY

The Lucky Louie property (fig. 11) is on the southern end of the low hills that form the south end of Spors Mountain.   It is in the NW¼ sec. 10, T. 13 S., R. 12 W., Salt Lake principal meridian.   The property is connected to the haulage road along the west side of the mountain by a narrow dirt road, several hundred yards long.   The property was staked by James Quigley, H. E. Carlson, E. H. Snell, and Hyram Schmidt of Eureka, Utah.   Little prospecting has been undertaken, though two widely separated bulldozer trenches have been made.   One trench exposed fluorite pebbles in a conglomerate that formed in Lake Bonneville.   The second trench exposed fluorspar, but the size and shape of the deposit are not known.   The minimum dimensions of this body appear to be 10 by 20 feet, but further exploration might expose a body many times this size.   James Quigley reports (personal communication) that a sample of this fluorspar contained 83 percent CaF_2.

NONELLA NO. 1 CLAIM

The Nonella No. 1 claim is on the eastern flank of Spors Mountain at an altitude of about 5,500 feet (fig. 11).   It is in the south-central part of T. 12 S., R. 12 W., Salt Lake principal meridian.   The property is 500 feet above the haulage road and is connected to it by a steep trail.   The Nonella No. 1 claim was located on April 20, 1948 by Nona Chesley and Ella Black of Delta, Utah.   Little development work has been done on this property.   A pit 15 feet long, 8 feet wide, and 10 feet deep has been dug on a fluorspar body in dark-gray dolomite that strikes N. 35° E. and dips 34° NW. The fluorspar body is covered almost entirely by talus made up of dolomite and chert boulders. White to pale-purple, fine-grained fluorite is exposed in the pit. Locally the fluorspar contains many thin stringers of opal. Fragments of altered rhyolite porphyry were found on the dump, but none was observed in place. The altered rhyolite porphyry is deep red and contains abundant phenocrysts of smoky quartz in an aphanitic groundmass of clayey, calcareous, limonitic material.

OVERSIGHT NO. 4 PROPERTY

The Oversight No. 4 property, the northernmost fluorspar deposit (fig. 11) in Spors Mountain, is in the southwest part of T. 12 S., R. 12 W., Salt Lake principal meridian. It is near the crest of a jagged hill that rises 800 feet above the valley floor. A rough dirt road, 1 mile long, leads from the main haulage road to the compressor house, and a trail leads southward to the mine workings. The claim was located in 1949 by Fred Staats, Frank Lowder, and Steve Stephenson.

The Oversight No. 4 property is developed by a 20-foot opencut and several prospect pits and bulldozer trenches. The opencut is on a vertical fluorspar pipe about 20 feet in diameter in a dense, fine-grained, black dolomite that strikes N. 34° E. and dips 32° NW. The ore consists of a boxwork of brownish fluorspar.

THURSDAY NO. 3 PROPERTY

The Thursday No. 3 property (fig. 11) is in the south-central part of T. 12 S., R. 12 W., Salt Lake principal meridian. An 8-mile dirt road along the western edge of Spors Mountain connects this mine to the main haulage route to Delta, Utah. The Thursday No. 3 claim was located September 5, 1948, by W. L. Hamblin, G. D. Hamblin, and A. W. Hamblin of Kaysville, Utah. Production records are not complete, although it was reported that about 100 tons of fluorspar has been produced from this claim. Mining was not in progress in September 1950.

The workings on the Thursday No. 3 property (fig. 17) consist of two adits, four opencuts, and several prospect pits. The main adit was driven S. 43° E. along a fault for 175 feet. The deposit at the main adit is developed on the surface by three opencuts, from which most of the fluorspar has been mined. The second adit, not shown on figure 17, was driven N. 80° W. for 300 feet from the eastern side of the hill, about 150 feet below the main adit level. The second adit intersects an 8- to 12-inch fluorspar vein at the face.

The fluorspar occurs as thin veins and replacement bodies in dolomite (fig. 17) that strikes N. 42° E. and dips 30°-34° NW. The

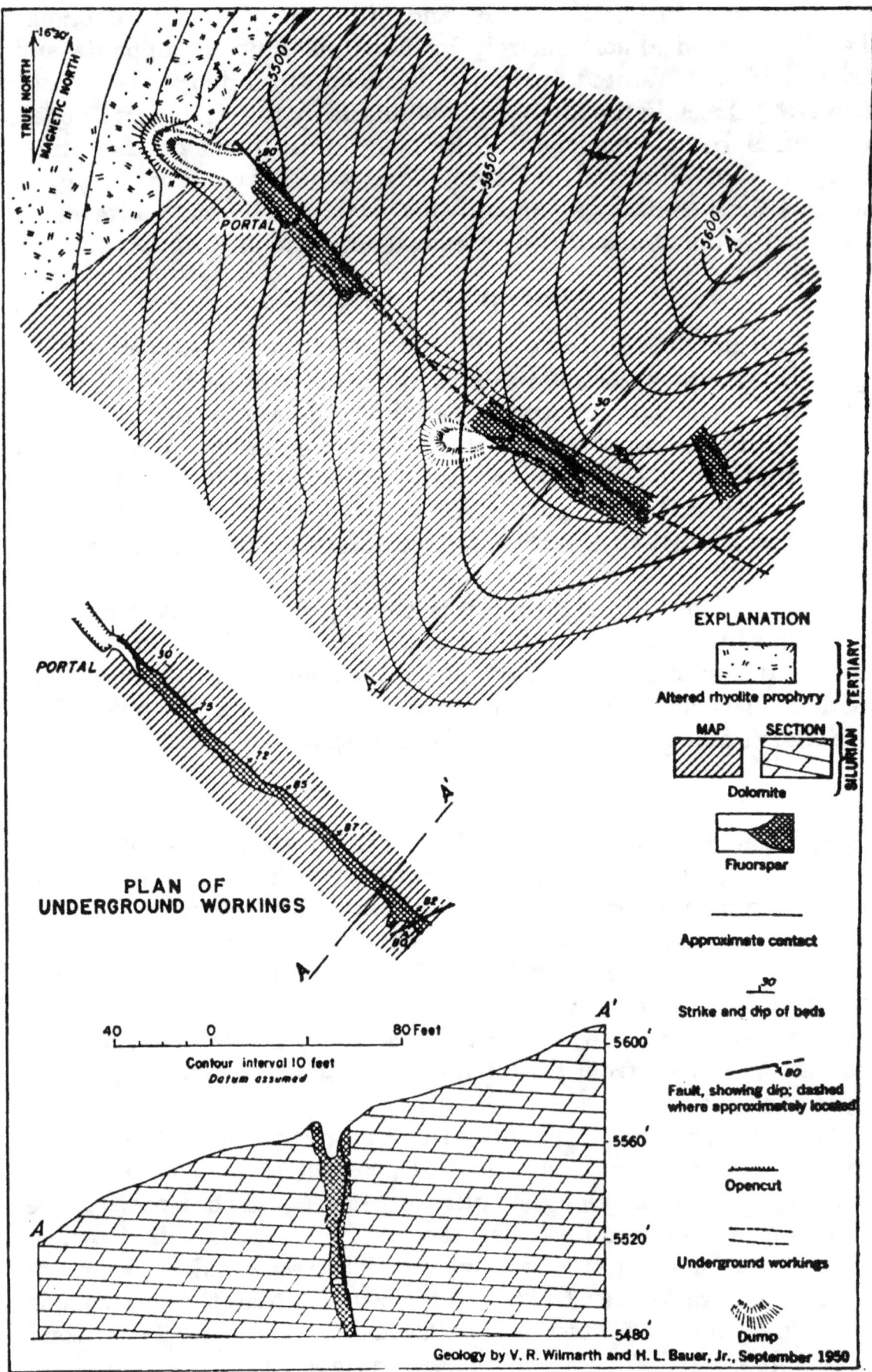

FIGURE 17.—Maps and section, Thursday No. 3 property, Juab County, Utah.

dolomite is intruded by a dike of highly altered, reddish-brown rhyolite porphyry. This rock contains orthoclase phenocrysts, altered to a greenish clay, in an aphanitic groundmass of quartz, feldspar, and dark-purple fluorite. In some places the rhyolite porphyry has a purplish hue because of the abundance of the fluorite. The dike trends northeast and was traced for several hundred feet along strike.

The major fault in the vicinity of the mine strikes N. 42°–60° W. and dips 72°–87° NE. It is exposed throughout the main adit and in the largest opencut, southeast of the adit portal. The amount and direction of movement on the fault could not be determined, though the displacement appears to be small. The dolomite has been brecciated on the footwall side of the fault.

Two fluorspar bodies are exposed in the Thursday mine workings. The largest, in the footwall of the fault, has a maximum thickness of 14 feet in the opencuts and 6 feet on the main adit level, 65 feet below the surface. This body has a minimum length of about 240 feet. A second fluorspar vein, exposed in a prospect pit about 50 feet east of the largest opencut, has a maximum length of 60 feet and a maximum width of 10 feet. The fluorspar in both bodies is white to pale purple, fine-grained and sugary, and locally contains small pockets of coarsely crystalline fluorite. Narrow veins of white crystalline calcite are numerous at the contact of the dolomite and the fluorspar.

SILVER QUEEN DEPOSIT, TOOELE COUNTY

By J. F. Smith, Jr., and A. H. Wadsworth, Jr.

The Silver Queen fluorspar deposit, Tooele County, is about 21 miles south of Clive, a station on the Western Pacific Railroad, which is 48 miles east of Wendover, Utah, on U. S. Highways 40 and 50. About 1,200 tons of high-grade fluorspar was mined at the Silver Queen deposit, also known as the Wildcat mine, between 1918 and 1924. In 1938 a few tons of fluorspar was mined.

The deposit was described by Heikes (1922), Buranek (1942), and Burchard (1933, p. 21). Buranek (1942, p. 21) reports that a lamprophyre dike occurs close to the fluorspar deposits. In his discussion of the age of the limestone in which the fluorspar occurs, Buranek (1942, p. 21) quotes a private report by R. E. Marsell, who interpreted the fossils to be of Carboniferous age. According to Burchard (1933, p. 21), at various places over a 90-acre tract surrounding the Silver Queen deposit, veins along faults of small displacement contain fluorite, barite, pyrite, chalcopyrite, and traces of gold and silver.

The fluorite veins at the Silver Queen deposit are on a limestone hogback trending north-south and on a low, flat limestone hill about

800 feet east of the hogback (fig. 18). The two hills rise above the deposits of former Lake Bonneville. On the hogback the limestone beds in general strike north to northwest and dip 20°–50° W. On the low hill they strike nearly east and dip 30° S. The reason for this difference in strike in the two areas was not determined during the short time spent on the property, but it may be the result of faulting.

The fluorite occurs as well-formed cubes that range from almost colorless to blue and green. Green copper staining and crystals of white witherite and dolomite are associated with the fluorite in the veins. A thin streak of reddish opaline material along the east side of the vein in the southern cut on the hogback was the only siliceous material observed.

The fluorite veins trend N. 75°–80° E. and N. 5°–35° E. Along the fissures the fluorite occurs in pockets, 5 to 15 feet long and 2 inches to 5 feet thick. The northward-trending vein on the hogback has a maximum thickness of 2 feet, pinches to 2 inches in places, and is not continuously mineralized with fluorite for the entire length indicated on the map (fig. 18).

Most of the fluorite has been removed from the trenches shown on the eastern hill, although small pockets of fluorite were left on the hanging walls of small faults.

The southernmost trenches that cross the eastern hill appear to be along a main fracture that trends N. 80° E. and dips 50° N. On the hogback a mineralized fault with this trend has a throw of 10 to 15 feet. The area between these outcrops is covered by the lake deposits, and a caved shaft about 10 feet deep along the projected trench contains only lake sediments.

A drift at a depth of 40 feet in the shaft on the east side of the low hill has been driven S. 35° W. along a 2-foot vein of fluorite on the east wall. This vein is traceable for about 100 feet, although it is pockety in places and ranges in thickness from a few inches to 2 feet. At the same depth a short drift trending S. 80° W. has cut a few small pockets of fluorite. At the bottom of the shaft, about 80 feet deep, small pockets of fluorite remain along the hanging wall of a fault striking N. 45° E. and dipping 55° NW. A short stope has been worked along this mineralized zone.

The northernmost inclined shaft on the hogback cuts a fluorite vein at the opening but follows another vein downward. The latter vein carries mostly dolomite and witherite. The lower inclined shaft to the southeast did not cut the northward-trending vein exposed higher on the hogback.

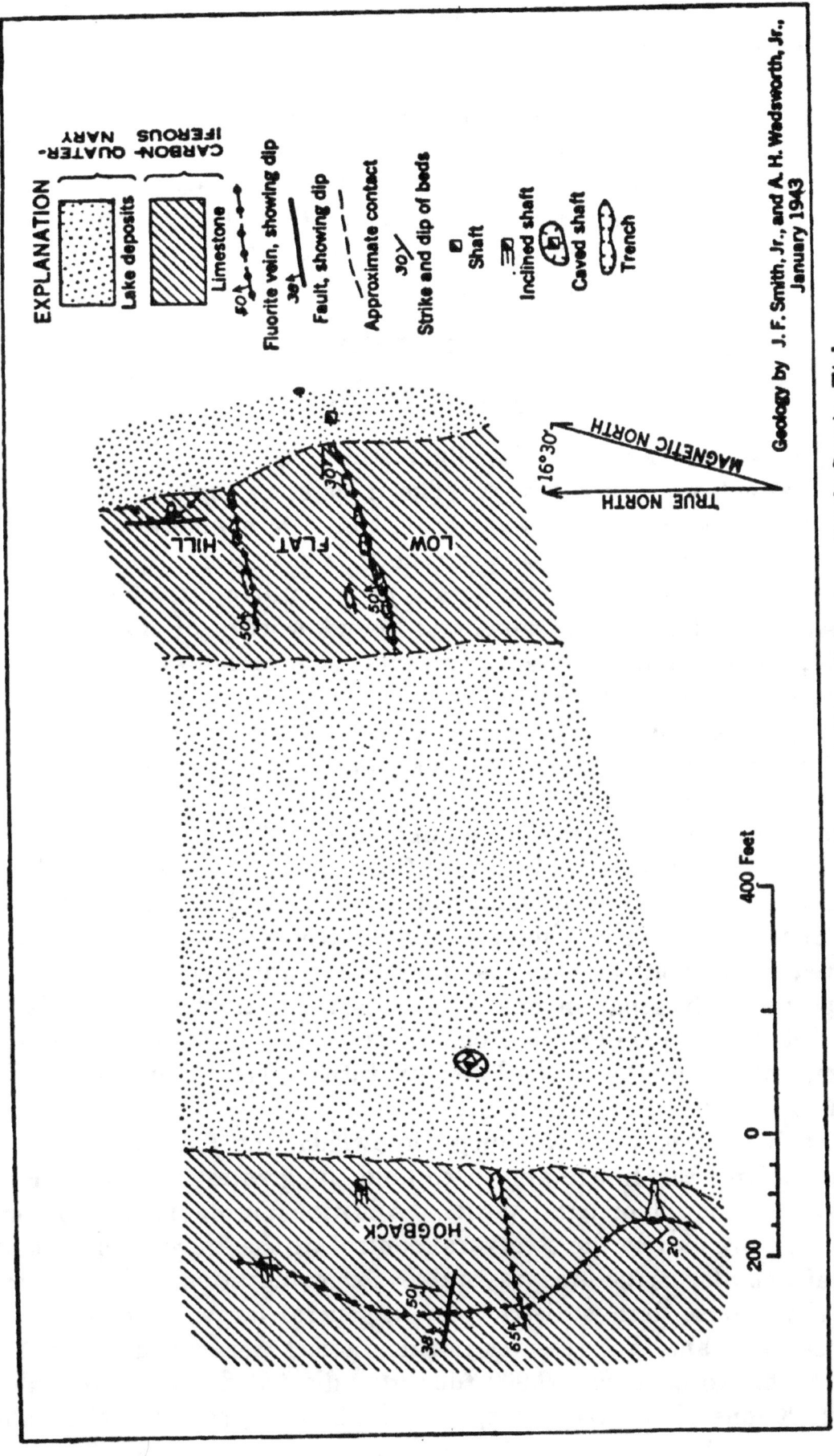

FIGURE 18.—Geologic sketch map of Silver Queen area, Tooele County, Utah.

On the hogback the fluorite does not occur above a quartzite bed that overlies the limestone. Several prospects in another hill several hundred feet west of the hogback failed to show any fluorite.

All the fluorite observed is in small pockets along the walls of the old workings and would have to be selectively mined. The aline-ment of the veins on the two hills suggests that they may continue across the intervening sand-covered area. Trenching is recommended for prospecting between the exposures on the hills; the thickness of the overburden, however, is unknown.

RESOURCES AND POSSIBLE DEVELOPMENT

By W. R. Thurston

The possibilities of mining fluorspar in Utah have not attracted wide attention until recently. The interest in the metallic ores of the State has overshadowed the less-known nonmetallic minerals, and it is doubtful that many prospectors have been aware of either the value of fluorspar or the means of recognizing it. Only one de-posit, the Silver Queen mine, was found worthy of mining before 1920, but since 1935 minable deposits have been located in widely separated parts of the State.

The critical need for fluorspar in World War II aroused interest in fluorspar mining and showed the need for intensive search for new deposits. War-born shortages of men and materials, however, made thorough study and exploration impossible. The result is that known deposits were greatly depleted and few new deposits have been found.

The known reserves of fluorspar in Utah total about 450,000 tons of material having a minimum of 40 percent of CaF_2. When com-pared with fluorspar deposits elsewhere in the Nation, most of the deposits of fluorspar in Utah appear small, but in part this is due to lack of adequate exploration and study. No estimate is made of reserves that might be disclosed by future geologic or exploratory work.

At the present stage of geologic knowledge the Thomas Range and the Indian Peak Range appear to hold the greatest promise for dis-coveries of new fluorspar deposits. Additional geologic mapping may reveal more clearly the structural control of the known deposits and locate similar structures favorable for exploration. The deposits in the Thomas Range district are in general larger than those else-where in Utah. The reserves in this district were estimated in 1950 by Staatz to be about 62,000 tons of indicated fluorspar and about 300,000 tons of inferred fluorspar. The known reserves in the Indian

Peak Range, at the Cougar Spar, Blue Bell, and JB mines, amount to about 50,000 tons of fluorspar containing approximately 40 percent of CaF_2. The crude fluorspar, however, contains abundant quartz and brecciated country rock and as a consequence is very siliceous. Simple gravity methods of concentration have not been entirely satisfactory for reducing the silica content, but the use of flotation methods of beneficiation probably would result in the recovery of high-grade fluorspar concentrates.

The Monarch (Staats) area of Beaver County can be expected to produce fluorspar for some time under present economic conditions. Washing equipment that the operator has installed at Lund, Utah, should make it profitable to mine and beneficiate the more clayey fluorspar that has been of submarginal grade in the past. Cross-faulted parts of the contact between the rhyolite porphyry and limestone should be sought and explored. Any new deposits probably will be high in fluorspar content, relatively low in silica content, and small, much like the known deposits.

The Thomas Range district in Juab County has many exposures of fluorspar, and 12 deposits have been mined successfully (table 2). Further study of the area may reveal the structural controls and guide the exploration for other fluorspar deposits. Many favorable localities remain for the discovery of additional fluorspar deposits in the district.

The Star district in Beaver County is not so favorable geologically as the Thomas Range district. Undoubtedly most of the known pockets of fluorspar have been mined, and underground exploration for additional deposits of such small size is not warranted. The recovery of byproduct fluorspar from the mining of metal ores in the Star district is dependent on the production of those ores.

The Silver Queen (Wildcat) area is even less favorable for further exploration. Since the period 1918 to 1924, when more than 1,000 tons of fluorspar was produced at the Silver Queen mine, no important bodies of fluorspar have been found. Additional fluorspar deposits may exist beneath the overburden, but unless they are larger than those mined heretofore they would be too difficult and costly to find.

Fluorite has been reported from other places in Utah. Whether any of these localities will become important producers of fluorspar is questionable, but in the absence of more nearly complete data some, at least, should be examined critically. The occurrences are listed in table 3 and shown in figure 1.

TABLE 3.—*Minor occurrences of fluorite in Utah*

County	Name of occurrence	Locality data	Fluorite occurrence	Source of data
Beaver	Cupric Mine Co. tungsten deposit, San Francisco district.	On west side of San Francisco Mts., 16 miles west of Milford.	Fluorite is one of the minor contact minerals developed in the scheelite-bearing sulfide ore body at the contact between sedimentary rocks and quartz monzonite.	Hobbs, S. W., 1944, Tungsten deposits in Beaver County, Utah: U. S. Geol. Survey Bull. 945-D, p. 89.
Beaver	Granite district, Bismuth mine and Big Pass and 2 R's prospects.	Mineral Range east of Milford	Fluorite is one of the minor contact minerals developed in limestone near the contact with granite; bismuthinite, molybdenite, and scheelite are among the associated minerals.	Butler, B. S., and others, 1920, The ore deposits of Utah: U. S. Geol. Survey Prof. Paper 111, p. 534. Crawford, A. L., and Buranek, A. M., 1945, Tungsten deposits of the Mineral Range, Beaver County, Utah: Utah Univ. Bull., v. 35, no. 15, p. 29, 46.
Beaver	Mystery and Snifter claims, Indian Creek.	North side of Indian Canyon, sec. 28, T. 27 S., R. 6 W., Marysvale region.	Fluorite, associated with autunite, occurs in Tertiary volcanic rocks.	Wyant, D. G., and others, 1950, Uranium resources in Marysvale region, Utah, an interim report: U. S. Geol. Survey Trace Elements Memo. Rept. 169.
Grand	Ryan Creek scheelite prospect of the late Joseph Quinn of Grand Junction, Colo.	In T. 22 S., R. 25 E.; at altitude of 5,200 ft., 1 mile east of the junction of Ryan and Cow Creeks, on Ryan Creek.	Fault breccia between sandstone and granite, cemented by fluorite-bearing vein material; associated minerals are carbonates and barite with traces of galena, sphalerite, and scheelite.	Field notes of Ogden Tweto and W. S. Burbank. In the files of the U. S. Geol. Survey. Also Dane, C. H., 1935, Geology of the Salt Valley anticline and adjacent areas, Grand County, Utah: U. S. Geol. Survey Bull. 863, p. 179.
Grand	Blue Spar prospect, W. B. Knight, Jr.	Near Grand Junction, Colo.	Not described	Davis, H. W., 1950, Fluorspar and cryolite: U. S. Bur. Mines Minerals Yearbook, 1948, p. 589.
Iron	Gold Springs district	17 miles northwest of Modena	Traces of fluorite common in the district; most conspicuous at prospect on summit of Bull Hill.	Butler, B. S., and others, 1920, The ore deposits of Utah: U. S. Geol. Survey Prof. Paper 111, p. 565.
Juab	Johnson Peak or Trout Creek district.	In Deep Creek Range, 20 miles south of Callao; on north side of Trout Creek near base of range.	Fluorite is conspicuous gangue mineral in dump at a prospect shaft sunk in search of sphalerite ore.	Butler, B. S., and others, 1920, The ore deposits of Utah: U. S. Geol. Survey Prof. Paper 111, p. 485.
Juab	Fish Springs district	T. 11 S., R. 14 W.; northwest end of Fish Springs Range, about 1 mile north of Joseph Tunnel and east of magnesite prospects.	Green fluorite in limestone; vein about 8 in. wide and 12 ft. long.	Buranek, A. M., 1946, Fluorite in Utah: Utah Dept. Publicity and Industrial Development, Circ. 36, p. 14.
Juab	West Tintic district, "68" mine.	About 21 miles southwest of Eureka; south margin of Sheeprock Mts.	Clear to purple fluorite is associated with quartz, barite, and galena.	Stringham, B. F., 1942, Mineralization in the West Tintic mining district, Utah: Geol. Soc. America Bull., v. 53, p. 285.
Millard	Rain Bow mine, Brice Frisby.	In sec. 30, T. 25 S., R. 6 W., about 1½ miles northeast of Cove Fort.	Not described	Davis, H. W., 1949, Fluorspar and cryolite: U. S. Bur. Mines Minerals Yearbook, 1947, p. 510.
Piute	Ohio district, Bully Boy and Webster mines.	About T. 27 S., R. 3 W.; in Tushar Range, Marysvale region.	Fluorite is a minor mineral in a quartz vein filling a fault in quartzite and lava; associated minerals are carbonates and barite.	Butler, B. S., and others, 1920, The ore deposits of Utah: U. S. Geol. Survey Prof. Paper 111, p. 556.

County	Locality	Description	Reference
Piute	A. Bullion Monarch (Farmer John) mine. B. Freedom and Prospector claims.	A. Near boundary between secs. 26 and 27, T. 26 S., R. 4 W. B. Sec. 4, T. 27 S., R. 3 W. Both in Antelope Range, Marysvale region.	A. White to purple fluorite occurs as fillings in steeply dipping fissures and breccia zones. B. The fluorite forms blebs and lenses in or adjacent to dense quartz veins in quartz monzonite.	Granger, H. O., and Bauer, H. L., Jr., 1950, "Preliminary examination of uranium deposits near Marysvile, Piute County, Utah: U. S. Geol. Survey Trace Elements Memo. Rept. 33, p. 23–27.
Summit	Park City district, Woodside mine and Silver King mine.	In Woodside Gulch, T. 2 S., R. 4 E.	Associated with uraniferous material at both localities. Fluorite is a conspicuous gangue mineral in ore from the Carey incline of the Woodside mine and is associated with oxidized replacement ore at the Silver King mine.	Boutwell, J. M., 1912, Ore deposits of the Park City district, Utah: U. S. Geol. Survey Prof. Paper 77, p. 187 and 190.
Tooele	Ophir district, Buffalo mine.	In Lion Hill area almost 2 miles up Long Trail Gulch from Ophir.	Fluorite is a gangue mineral of sulfide ore in limestone.	Gilluly, James, 1932, Ore deposits of Stockton and Fairfield quadrangles, Utah: U. S. Geol. Survey Prof. Paper 173, p. 139–140.
Tooele	Dugway district, Rattler No. 38 claim (Cannon prospect).	T. 10 S., R. 12 W., Dugway Mts.	Quartz veins in quartzite contain some fluorite.	Field notes of A. E. Granger in files of the U. S. Geol. Survey.
Tooele	Granite Mountain district, Desert claim.	Near north end of Granite Range.	Fluorite in a vein with quartz, hematite, and locally galena and chalcopyrite.	Buranek, A. M., 1948, Fluorite in Utah: Utah Dept Publicity and Industrial Development, Circ. 36, p. 15. Butler, B. S., and others, 1920, The ore deposits of Utah: U. S. Geol Survey Prof. Paper 111, p. 462.
Tooele	East Erickson district, Silver King and Copper Jack prospects.	In Sheeprock Mts.; west of Godiva claims; approximately in T. 9 S., R. 6 W.	Uraniferous sulfide-fluorite-quartz veins in granite.	Butler, B. S., and others, 1920, The ore deposits of Utah: U. S. Geol. Survey Prof. Paper 111, p. 427–428. Wyant, D. G., and others, 1950, Uranium resources in Marysvale region, Utah, an interim report: U. S. Geol. Survey Trace Elements Memo. Rept. 169, p. 51–52.
Tooele	West Erickson district.	In Simpson Mts.	Fluorite associated with leached quartz and lead carbonate, chiefly in quartzite.	Butler, B. S., and others, 1920, The ore deposits of Utah: U. S. Geol. Survey Prof. Paper 111, p. 452.
Tooele	Gold Hill district, Gold Bond claim.	In T. 8 S., R. 18 W.; in Lucy L Gulch about 1½ miles south of Gold Hill.	Small amounts of purple and white fluorite are associated with danburite in a vein in quartz monzonite.	Nolan, T. B., 1935, The Gold Hill mining district, Utah: U. S. Geol. Survey Prof. Paper 177, p. 112, 126.
Weber	Weber district.	In Cold Water Canyon, 2 miles east of Ogden.	Fluorite reported to occur with barite, quartz, and copper minerals in a vein between quartzite and limestone.	Butler, B. S., and others, 1920, The ore deposits of Utah: U. S. Geol. Survey Prof. Paper 111, p. 223.
Weber	Weber district.	2½ miles northeast of Ogden.	Narrow fluorite veins in gneiss and pegmatite of pre-Cambrian age.	Burchard, E. F., 1933, Fluorspar deposits in western United States: Am. Inst. Min. Met. Eng. Tech. Pub. 500, p. 21.

LITERATURE CITED

Boutwell, J. M., 1913, Ore deposits of the Park City district, Utah: U. S. Geol. Survey Prof. Paper 77.

Buranek, A. M., 1942, The fluorspar deposits of the Wild Cat Mountains, Tooele County, Utah: Utah Dept. Publicity and Industrial Devel., Circ. 5, p. 20-23.

———— 1948, Fluorite in Utah: Utah Dept. Publicity and Industrial Devel., Circ. 36, 25 p.

Burchard, E. F., 1933, Fluorspar deposits in western United States: Am. Inst. Min. Met. Eng. Tech. Pub. 500, p. 3-25.

Butler, B. S., 1913, Geology and ore deposits of the San Francisco and adjacent districts, Utah: U. S. Geol. Survey Prof. Paper 80.

Butler, B. S., Loughlin, G. F., Heikes, V. C., and others, 1920, The ore deposits of Utah: U. S. Geol. Survey Prof. Paper 111.

Crawford, A. L., and Buranek, A. M., 1945, Tungsten deposits of the Mineral Range, Beaver County, Utah: Utah University Bull., v. 35, p. 29, 46.

Dane, C. H., 1935, Geology of the Salt Valley anticline and adjacent areas, Grand County, Utah: U. S. Geol. Survey Bull. 863.

Davis, H. W., 1950, Fluorspar and cryolite: U. S. Bur. Mines Minerals Yearbook. 1948, p. 589.

Everett, F. D., and Wilson, S. R., 1950, Investigation of the J. B. fluorite deposit, Beaver County, Utah: U. S. Bur. Mines Rept. Inv. 4726, 11 p.

———— 1951, Investigation of the Cougar Spar fluorspar deposit, Beaver County, Utah: U. S. Bur. Mines Rept. Inv. 4820, 12 p.

Fitch, G. A., Quigley, James, and Barker, C. S., 1949, Utah's new mining district: Eng. and Min. Jour., v. 150, p. 63-66.

Gilluly, James, 1932, Ore deposits of Stockton and Fairfield quadrangles, Utah: U. S. Geol. Survey Prof. Paper 173.

Heikes, V. C., 1922, A fluorspar deposit in Utah: Mineral Resources U. S., 1921, pt. 2, p. 48-49.

Hobbs, S. W., 1944, Tungsten deposits in Beaver County, Utah: U. S. Geol. Survey Bull. 945-D.

Marsh, J. A., and Everett, F. D., 1945, A new domestic source of fluorspar in Utah: Eng. and Min. Jour., v. 146, p. 98-100.

Muessig, Siegfried, 1951, Eocene volcanism in central Utah: Science, v. 114, p. 234.

Nolan, T. B., 1935, The Gold Hill mining district, Utah: U. S. Geol. Survey Prof. Paper 177.

Stringham, B. F., 1942, Mineralization in the West Tintic mining district, Utah: Geol. Soc. America Bull., v. 53, p. 285.

INDEX

www.ingramcontent.com/pod-product-compliance
Lightning Source LLC
Chambersburg PA
CBHW081900170526
45167CB00007B/3091